Are Cell Phones Dangerous?

Other titles in the *In Controversy* series:

Are Video Games Harmful?
Can Renewable Energy Replace Fossil Fuels?
Childhood Obesity
Does the Death Penalty Deter Crime?
Does Illegal Immigration Harm Society?
How Dangerous Are Performance-Enhancing Drugs?
How Serious a Threat Is Climate Change?
How Should America Respond to Illegal Immigration?
How Should the World Respond to Global Warming?
Is Animal Experimentation Ethical?
Is Offshore Oil Drilling Worth the Risks?
Is Stem Cell Research Necessary?
Is the World Prepared for a Deadly Influenza Pandemic?
Should the Drinking Age Be Lowered?
Should Juveniles Be Tried as Adults?
Should Marijuana Be Legalized?

Are Cell Phones Dangerous?

Bonnie Szumski and Jill Karson

IN CONTROVERSY

San Diego, CA

Printed in the United States

For more information, contact:
ReferencePoint Press, Inc.
PO Box 27779
San Diego, CA 92198
www. ReferencePointPress.com

LIBRARY OF CONGRESS CATALOGING-IN-PUBLICATION DATA

Szumski, Bonnie, 1958-
Are cell phones dangerous? / by Bonnie Szumski and Jill Karson.
p. cm. -- (In controversy series)
Includes bibliographical references and index.
ISBN-13: 978-1-60152-232-0 (hardback)
ISBN-10: 1-60152-232-0 (hardback)
1. Cell phones--Health aspects. I. Karson, Jill. II. Title.
RA569.3.S98 2012
615.9'25663--dc23
2011034063

Contents

Foreword

In 2008, as the US economy and economies worldwide were falling into the worst recession since the Great Depression, most Americans had difficulty comprehending the complexity, magnitude, and scope of what was happening. As is often the case with a complex, controversial issue such as this historic global economic recession, looking at the problem as a whole can be overwhelming and often does not lead to understanding. One way to better comprehend such a large issue or event is to break it into smaller parts. The intricacies of global economic recession may be difficult to understand, but one can gain insight by instead beginning with an individual contributing factor, such as the real estate market. When examined through a narrower lens, complex issues become clearer and easier to evaluate.

This is the idea behind ReferencePoint Press's *In Controversy* series. The series examines the complex, controversial issues of the day by breaking them into smaller pieces. Rather than looking at the stem cell research debate as a whole, a title would examine an important aspect of the debate such as *Is Stem Cell Research Necessary?* or *Is Embryonic Stem Cell Research Ethical?* By studying the central issues of the debate individually, researchers gain a more solid and focused understanding of the topic as a whole.

Each book in the series provides a clear, insightful discussion of the issues, integrating facts and a variety of contrasting opinions for a solid, balanced perspective. Personal accounts and direct quotes from academic and professional experts, advocacy groups, politicians, and others enhance the narrative. Sidebars add depth to the discussion by expanding on important ideas and events. For quick reference, a list of key facts concludes every chapter. Source notes, an annotated organizations list, bibliography, and index provide student researchers with additional tools for papers and class discussion.

The *In Controversy* series also challenges students to think critically about issues, to improve their problem-solving skills, and to sharpen their ability to form educated opinions. As President Barack Obama stated in a March 2009 speech, success in the twenty-first century will not be measurable merely by students' ability to "fill in a bubble on a test but whether they possess 21st century skills like problem-solving and critical thinking and entrepreneurship and creativity." Those who possess these skills will have a strong foundation for whatever lies ahead.

No one can know for certain what sort of world awaits today's students. What we can assume, however, is that those who are inquisitive about a wide range of issues; open-minded to divergent views; aware of bias and opinion; and able to reason, reflect, and reconsider will be best prepared for the future. As the international development organization Oxfam notes, "Today's young people will grow up to be the citizens of the future: but what that future holds for them is uncertain. We can be quite confident, however, that they will be faced with decisions about a wide range of issues on which people have differing, contradictory views. If they are to develop as global citizens all young people should have the opportunity to engage with these controversial issues."

In Controversy helps today's students better prepare for tomorrow. An understanding of the complex issues that drive our world and the ability to think critically about them are essential components of contributing, competing, and succeeding in the twenty-first century.

A Ubiquitous Technology

A relatively new invention, cell phones have entered American society—and the world—with a vengeance. People use cell phones everywhere, even when doing so poses a danger, such as while driving. As accidents rise, state and federal governments have passed numerous laws banning handheld cell phones, banning texting, or banning phone use altogether while driving.

The passage of these laws has had little effect, and in the case of texting, might even be counterproductive. It is not uncommon to see drivers looking down at their laps rather than at the road as they try to hide their texting from the eyes of passing police officers. Some studies have even shown that driving with a cell phone is more dangerous than driving drunk.

"Switching off their phones causes [teens] anxiety, irritability, sleep disorders, or sleeplessness, and even shivering and digestive problems."[1]

— Lopez Torrecillas, lecturer at the Department of Personality and Psychological Assessment and Treatment at the University of Granada, Spain.

The cell phone has impacted society in many ways beyond contributing to dangerous driving habits. Many people carry their cell phones with them at all times—checking messages continually even in the middle of meetings and conversations, posting updates on their latest shopping spree or meal and, in general, spending countless hours each day online on their phones. An archaeologist who studied teen behavior and cell phone use found that many teens will sit side-by-side, separately texting or talking on their phones rather than communicating with one another. These new behaviors have been redefining the social fabric of the United States and some other countries as well.

The cell phone does have many positive aspects, including facilitating communication between individuals, such as children and parents, enabling the access of information quickly, and, in general, acting as a mobile computer. Others use the cell phone to look up facts and information as they discuss an issue with another person, or use the phone's GPS to find their way to various locations.

The cell phone has become an essential tool of modern life: People talk and text on them, they look up addresses, watch movies, play games, check prices, take pictures, and more. Cell phones have also become a dangerous distraction on the road and they might pose other health hazards.

Addiction

However, constant cell phone use may be creating behavioral changes in users and influencing young adults in a particularly negative way. In light of certain health concerns, some experts even see the persistent use of the cell phone as an addiction. In one study in Spain, 40 percent of young adults were found to be using their cell phones more than four hours a day. Lopez Torrecillas, lecturer at the Department of Personality and Psychological Assessment and Treatment at the University of Granada,

Spain, claims that cell phone addiction is as bad as addiction to drugs, tobacco, or alcohol but that instead of physical problems, the phones cause psychological ones. "Switching off their phones causes [teens] anxiety, irritability, sleep disorders, or sleeplessness, and even shivering and digestive problems,"[1] Torrecillas claims.

The comparison between the physical addiction of drugs and alcohol and the psychological addiction of the cell phone may not be that far off. Many see the resistance to giving up the cell phone while driving to be similar to an addiction—even though people are eight times more likely to be in an accident while using a cell phone, the behavior persists.

The cell phone continues to be studied for psychological effects on the individual and society, for evidence that it may be sowing the seeds for health effects such as cancerous brain tumors, and to determine whether its use will foster negative, addictive behavior in young adults. Whether the cell phone is harmful to society and the individuals who use them will remain hotly debated as scientists and users attempt to answer these questions in the years ahead.

CHAPTER ONE

What Are the Origins of the Cell Phone Debate?

The cell phone has become an integral part of the everyday lives of billions of people across the globe. This small portable device can trace its roots back to 1946, when the first "portable" phone, the car phone, was introduced by AT&T. Built into cars, these phones were so bulky and expensive that they were typically employed only by the police, military, or wealthy individuals. Although these devices were cumbersome to use, the idea of a portable phone was hugely appealing to many people. As Guy Klemens describes it in his book *The Cellphone: The History and Technology of the Gadget That Changed the World*: "Post-war prosperity, and the technological progress achieved by then, allowed the creation of a new frivolity, the car phone. . . . Too large to be carried in anything less than an automobile and too expensive for any but the elite, the car phone nonetheless was popular. The systems were actually too popular, prompting improvements both in service quality and in capacity which would eventually lead to the cellphone."[2]

Birth of the First Truly Portable Phone

One of the key players in the telecommunications industry who was captivated by the car phone was Motorola senior executive and researcher Martin Cooper. In 1972 Cooper made a bold pronouncement that would dramatically expedite the development of cell phones: "What the world really needs is a handheld portable phone."[3] After a heated race against rivals at Bell Labs, Cooper made the first mobile phone call on April 3, 1973, demonstrating the first truly portable handheld telephone.

By the time the first mobile phone made its debut, many of the engineering technologies necessary for mass production had already been developed. In the most basic sense, cell phones are radio receivers and transmitters that use electromagnetic energy to send and receive large amounts of information wirelessly. This type of energy travels in waves, and the distance from the peak of one wave to the top of the next is referred to as the wavelength.

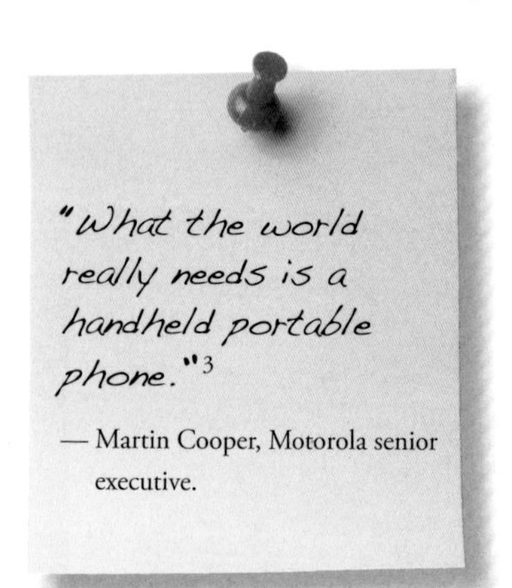
"What the world really needs is a handheld portable phone."[3]

— Martin Cooper, Motorola senior executive.

A related characteristic of waves is the frequency—that is, the number of times a wave oscillates up and down per second. The faster a wave oscillates, the higher its frequency and the more energy it carries. Frequency is measured in units called hertz (Hz), after physicist Heinrich Hertz, who in 1894 demonstrated that electromagnetic waves could be sent through space at the speed of light. A single hertz is one wavelength cycle per second. The electromagnetic waves that provide electricity to homes and businesses are powered with waves that operate at 60 Hz. Electronic devices that do not require wires operate at much higher frequencies, between 9,000 and 300 billion hertz. Wireless cell phones typically use radio waves in the 824 to 894 MHz frequency bands. MHz, or megahertz, refers to a million wavelength cycles per second. The most advanced generation of cell phones today operates at even higher power in the 2.1 GHz range. GHz, or gigahertz, represents a billion wavelength cycles per second.

The fastest-oscillating waves, such as those used in X-rays, are called ionizing radiation because they operate at such high frequencies that they can break the ionic bonds of atoms. When ionization occurs, electrons are stripped from atoms and molecules, which produces molecular changes that can damage biological tissue. Cell phones, on the other hand, use a weaker form of electromagnetic radiation in the nonionizing microwave range, which is not strong enough to break chemical bonds through ionization. This low-energy radiation is referred to as radio frequency (RF) energy or radiation.

Growth of the Industry

To work, a cell phone must send and receive millions of signals using electromagnetic waves. For example, a transmitter inside the phone turns a user's voice into invisible radio frequency signals that are sent out from the cell phone's antenna. Each cell phone must be able to transmit these signals over wide geographical areas. Broadcast areas are called "cells," and each cell comprises a tower with radio equipment. The towers broadcast the cell phone signals and allow callers to connect with other cell phones in the area. The towers transmit signals from cell to cell, forming a network of information-carrying radio waves. When a call is placed, the cell phone's antenna sends a signal to the nearest base station antenna. From there, the call is routed through a switching center, where the call can be picked up by another cell phone, another base station, or a land-based phone system.

The early car phone (pictured) was large, clunky, expensive, and limited in what it could do. It bears little resemblance to today's small, sleek, versatile cell phones.

By the late 1980s, a network of cell phone systems ran around the world. But it was not until the digital revolution of the 1990s that cell phones became common. Digitization provided a compressed way of sending and receiving information that made cell phone connections clearer and faster. With the availability of high-tech cell phones and the networks to support them, cellular communication exploded across the globe.

Cellular Towers

In the past two decades, cellular towers have become a permanent feature of many urban and rural landscapes. Because the radiation emitted from cellular towers is continuous and more powerful than radiation emitted from cell phones, concerns surfaced early that living or working near these structures could cause cancer and other health problems.

The cellular towers that house antennas and other equipment that support cellular telephony are typically 50 to 200 feet (15.2 m to 61 m) high. The energy from the cellular antenna is directed outward toward the horizon, parallel to the ground, with a small amount of downward scatter. The amount of energy decreases exponentially with increasing distance from the antenna. For this reason, many experts believe that ground exposure to RF radiation is quite low compared with the level close to the antenna. According to the Federal Communications Commission, "Exposure levels on the ground . . . are typically thousands of times below safety limits. . . . Therefore, there is no reason to believe that such towers could constitute a potential health hazard to nearby residents." Nevertheless, base station safety will likely be an area of active research for years to come.

Federal Communications Commission, "Radiofrequency Safety," 2011. www.transition.fcc.gov.

Over the next two decades, the industry grew at a stunning pace. Between 1990 and 2010, global mobile phone subscriptions grew from 12.4 million to nearly 5 billion. Unlike many other technological advances that benefit only the most affluent societies, the cell phone has become an integral part of daily life in both the developed and the developing world. By 2004, for example, as many as 75 million subscribers in Africa had adopted the technology.

Klemens describes how the cell phone permeates even the most unlikely regions of the globe:

> Many areas of the Congo River probably meet the definition of an undeveloped region. There is little electricity or plumbing, and certainly few telephone lines. In 2005 an American reporter found a woman selling fish who kept them in a river since there was no refrigeration, taking them to customers when requested. She cannot read or write, so sending a note will not help. Someone [who] wants to buy fish will just have to call her cellphone."[4] This is why, Klemens continues, the "cellphone is too vital for developing countries not to adopt. For many in the world, the cellphone brings a flowering of opportunity, with the century of technology finally making drastic improvements to their daily lives.[5]

Multipurpose Devices

Unlike the telephones of old, the cell phone has evolved far beyond its original purpose, that is, making and receiving calls. Today, cell phones support a variety of other services, including text messaging, gaming applications, music, and photography. Cell phones that contain minicomputers that enable them to support Internet access and other advanced applications are referred to as smart phones. Cell phones have indeed changed the world. As Devra Davis, author of *Disconnect: The Truth About Cell Phone Radiation, What the Industry Has Done to Hide It, and How to Protect Your Family*, puts it:

> Cell phones have revolutionized our ability to respond to emergencies real and imagined. They provide social status. They help you find a job. They neatly store all the music you want to listen to and summon up loved ones anytime,

anywhere. They seem essential to connecting you with whatever you want to do: track stocks, share favorite photos, download sports results, and send text, images, videos, and voice notes throughout the global village. To succeed you need to be reachable 24/7. And who can argue that these wonderful gadgets do make you feel more in touch, more effective, more, well, fun? The newest generation of phones brings us closer, faster, and nearer to one another than ever before. Cell phones today are like electricity and water—things we can't live without.[6]

Safety Questions Emerge

When Motorola introduced the first cell phone model in the early seventies, the company believed that the device's radiofrequency radiation was completely harmless. As cell phones became widely used in the United States in the 1990s, however, concerns about whether the radiation from cell phones posed a threat to human health emerged. The possibility was first proposed on the *Larry King Show* in 1993. King's featured guest, David Reynard, had lost his wife, Susan Reynard, the year before to a rare form of brain cancer. Reynard was suing his cell phone company, NEC, alleging that his wife's cancer was the result of her frequent cell phone use. To Reynard, the shape of his wife's brain tumor, a line swerving from the left side of her midbrain to her hindbrain, eerily matched the exact pattern of Susan's cell phone, which she held against her left ear for hours each day. The tumor's uncanny resemblance to a cell phone, and the fact that it developed in the exact location where his deceased wife had held her phone, convinced Reynard that radiation from the cell phone caused Susan's cancer and ultimate death.

Reynard's lawsuit was rejected in 1995 on the grounds that there was insufficient evidence to link cell phone radiation and brain cancer. However, the case prompted a slew of congressional hearings and scientific studies examining the safety of cell phones.

> "The newest generation of phones brings us closer, faster, and nearer to one another than ever before. Cell phones today are like electricity and water—things we can't live without."[6]
>
> — Devra Davis, founder of the Environmental Health Trust and author of *Disconnect: The Truth About Cell Phone Radiation, What the Industry Has Done to Hide It, and How to Protect Your Family.*

Martin Cooper holds a prototype of the first hand-held cellular telephone. Cooper used such a phone in April 1973 to make the first call from a mobile phone.

Early Research

Because cell phones are usually held close to the head, and the tissues next to where the phone is held can absorb radiation, much of the ensuing research focused on whether cell phone radiation could increase the risk of brain cancer. One of the earliest studies to examine this issue followed more than 420,000 adult cell phone users in Denmark between 1982 and 2002. The 20-year study did not establish a link between cell phone use and an increased risk of brain tumors.

Other early reports in the scientific literature reached different conclusions, however. Henry Lai, a bioengineering research professor at the University of Washington, first began conducting animal studies involving exposure to radiation as early as 1980. Lai found that rats exposed to radiofrequency radiation similar to that associated with wireless communication showed evidence of damaged DNA, which can lead to cancer and other ailments. A similar study by Lai and his colleague Narendra Singh published in a 1995 issue of *Bioelectromagnetics* also found damaged DNA in the brain cells of rats after a two-hour exposure to microwave radiation. The evidence remains ambiguous, however, as further research has failed to replicate these negative effects.

While these conflicting findings offered no definitive conclusions to put the controversy to rest, they suggested that the safety of cell phones was uncertain and that further study to determine the validity of these claims was warranted. With the number of cell phone users increasing rapidly in the 1990s, many of the governmental organizations that oversee public health stepped forward to monitor scientific findings to determine whether regulations were needed to protect human health. In 1997, for example, the US Food and Drug Administration (FDA), which regulates devices that emit radiation, including cell phones, stated in a letter to Congress that "little is known about the possible health effects of repeated or long-term exposure to low levels of radio frequency radiation (RFR) of the types emitted by wireless communications devices."[7]

In 2000 the FDA advised the National Toxicology Program that "there is currently insufficient scientific basis for concluding either that wireless communication technologies are safe or that they pose a risk to millions of users. A significant research effort, involving large, well-planned animal experiments is needed to provide the basis to assess the risk to human health of wireless communications devices."[8] In recent years, the FDA has stated that studies reporting biological effects linked to cell phone radiation cannot be replicated and therefore are inconclusive.

The National Cancer Institute has taken a similar stance, stating that

> most studies to date have not found an association between cell phone use and development of tumors. However, results from these studies have been limited by the

> length of follow-up, changing patterns of cell phone usage and technology, lack of study of children, and methods for measuring cell phone use. Possible cancer risks of cell phone exposure should continue to be evaluated using high-quality methodological approaches, particularly in relation to use in childhood and adolescence and longer-term use.[9]

The Advent of Safety Standards

As a result of the enormous increase in cell phone usage in the 1990s, a number of health and governmental authorities in many countries started to adopt standards to ensure public safety. In the United States, safety standards were set by the Federal Communications Commission (FCC) in 1996. These standards focused on limiting exposure to the radiation that is absorbed into the head and other body parts during cell phone use because this is the single biological effect of cell phone use that can be directly measured. The unit of measurement to quantify these levels is the specific absorption rate, or SAR. According to the FCC mandates set in the 1990s, the SAR for all phones sold in the United States should be no more than 1.6 watts per kilogram.

"Possible cancer risks of cell phone exposure should continue to be evaluated . . . particularly in relation to use in childhood and adolescence and longer-term use."[9]

— National Cancer Institute.

In codifying these standards, scientists used a nonhuman model, the Standard Anthropomorphic Man (SAM), a hypothetical six-foot-tall man weighing 200 pounds with a plastic head filled with liquid. Using SAM, scientists measured how much radiofrequency radiation was absorbed into the head when a cell phone was held a third of an inch from SAM's ear. Some are concerned that these standards, set in the nineties, need to be revamped to represent the typical cell phone user of today. As Davis puts it:

> SAM is not an ordinary guy. He ranked in size and mass at the top 10 percent of all military recruits in 1989 [and] SAM was not especially talkative, as he was assumed to use a cell phone for no more than six minutes at a time. . . .

> Unlike the rest of us, SAM's brain was uniform throughout. . . . Our heads and those of our children, of course, are not one simple, consistent, gooey liquid but contain many different components, including the hypothalamus, amygdala, and bone marrow. But SAM is a simpler fellow from simpler days when the idea that toddlers could be using cell phones was unimaginable.[10]

SAR Measurements and Public Safety

Like Davis, others are skeptical that these safety regulations will completely protect the consumer. For example, while the rate of absorption is quantifiable, how SAR measurements correspond to real usage is variable; a cell phone user's level of exposure depends on many factors. Different phone types and how they are held to the head cause variations in these measurements, for example. The farther a cell phone is from the nearest base station, the more power is required to initiate and maintain a cellular connection, which may increase exposure. The overall quality of the connection, too, affects a cell phone user's level of exposure.

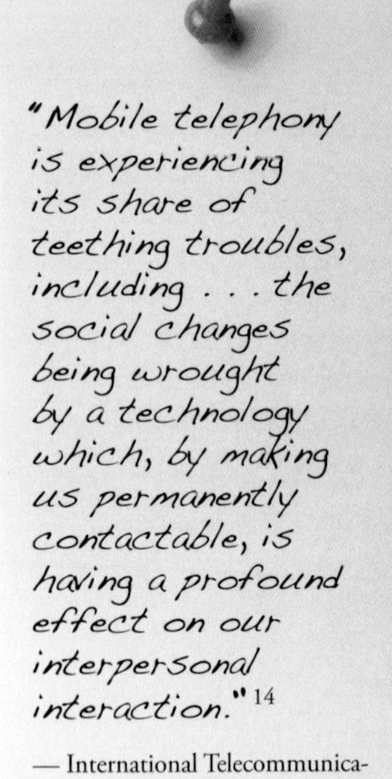
"Mobile telephony is experiencing its share of teething troubles, including . . . the social changes being wrought by a technology which, by making us permanently contactable, is having a profound effect on our interpersonal interaction."[14]

— International Telecommunication Union (ITU).

According to John Walls, the vice president of public affairs of CTIA, the international association for the wireless communications industry, these concerns are unfounded. Walls believes that SAR ratings ensure public safety: "What science tells us is, 'If the sign on the highway says safe clearance is 12 feet,' it doesn't matter if your vehicle is 4 feet, 6 feet or 10 feet tall; you're going to pass through safely. The same theory applies to SAR values and wireless devices."[11]

A scientific consensus has not been reached on whether current safety regulations are adequate. A particular concern that has emerged in recent years is how cell phones impact the health of children and teenagers. To date, no long-term studies on children and cell phone use have been completed, although children and teens represent one of the

A New Kind of Threat

Today, a third of all cell phone users in the United States own a smart phone. With so many individuals using these innovative devices, a new concern has emerged in recent years. At issue is whether smart phone users are vulnerable to hacker attacks.

A team of researchers at Rutgers University in New Jersey employed a particular type of malicious software called a rootkit to demonstrate how easily they could hack into smart phones. Rootkits have been used for years to infiltrate computer systems, and since today's smart phones are essentially mobile computers, they are just as vulnerable to these types of attacks. Because people carry smart phones around at all times, however, the social consequences of these attacks may be greater. The research team showed, for example, how the microphone on a smart phone could be turned on remotely using a rootkit, allowing an eavesdropper to listen in on conversation or anything else taking place around a user. With a simple text message, the team was also able to activate a user's GPS receiver to track the location of the owner. Hackers may even be able to access the personal information that is stored on a smart phone or to activate features to drain the battery and make the phone inoperable—all from a remote location. Whether or not this new danger can be counteracted remains to be seen.

fastest-growing populations of cell phone users and will have a potentially longer lifetime exposure to cell phone radiation. Some evidence suggests that children may be more vulnerable to the radiation emanating from cell phones than adults; radiation that penetrates two inches into an adult brain is absorbed much deeper in a child's brain because their skulls are thinner and their brains are more absorptive. As the search for more definitive answers continues, many experts are looking beyond health issues and assessing how these devices impact culture and social behavior.

Changing Cultural Norms

Today, the cell phone is one of the most dynamic technological tools of the Digital Age—and one that profoundly affects people of all ages and from all demographics. It is now a regular part of daily life and, for young people especially, appears to be the most popular form of electronic communication. The nearly 300 million cell phone subscriptions in the United States in 2010, translate into nearly one cellphone for every adult and child in the nation. In addition to the exponential increase in the number of cell phone users in recent years, the National Cancer Institute notes also that "over time, the number of cell phone calls per day, the length of each call, and the duration of use of cell phones have increased and cell phone technology has undergone substantial change."[12]

Gerard Goggin, professor of digital communication at the University of New South Wales, comments on how the ubiquity of cell phones impacts culture in his book *Cell Phone Culture: Mobile Technology in Everyday Life:*

> A bewildering and proliferating range of cultural activities revolve around cell phones: staying in constant contact, text messaging, fashion, identity-construction, music, mundane daily work routines, remote parenting, interacting with television programs, watching video, surfing the Internet, meeting new people, dating, flirting, loving, bullying, mobile commerce, and locating people. . . . They fit into new ways of being oneself (or constructing identity and belonging to a group); new ways of organizing and conducting one's life; new ways of keeping in touch with friends, romantic intimates, and family; new ways of conducting business; new ways of accessing services or education.[13]

Potential Drawbacks

Most would agree that the ability to stay connected with friends and family, send and receive messages, access the Internet, store

Modern cell phones (pictured) are not just telephones. They are minicomputers that essentially perform all of the functions of larger computers but packaged in a small, handheld device.

and listen to music, and take photos and videos—anywhere and anytime—facilitates many aspects of modern life. At the same time, the rapid development of cellular communication over the past 15 years has caused concern about the possibility of unanticipated consequences that have not been fully explored. As the International Telecommunication Union (ITU), the United Nations agency that specializes in communication and information technologies, states: "Like most young technologies, mobile telephony is experiencing its share of teething troubles, including . . . the social changes being wrought by a technology which, by making us permanently contactable, is having a profound effect on our interpersonal interaction."[14]

Of particular concern is that more and more people are using cell phones and texting as opposed to engaging in face-to-face contact and conversation. Communication via cell phones lacks nonverbal cues, such as facial expressions and gestures. Without these, communication may be less personal and more open to misinterpretation. Another concern is that people may become so engrossed with texting or any of the vast array of technologies available via the cell phone that they become oblivious to their surroundings, which may prevent them from making friends or participating in real-world activities with other people. The ITU calls it a trend that "threatens to eat away at our sense of social cohesion. Whether it's the novelty of the technology or our simple need to feel wanted, the human brain seems to register incoming electronic signals as inherently more urgent and important than the interpersonal signals coming from a fellow human being in front of us."[15]

These and other concerns are especially relevant when assessing how cell phones affect the behavior of young people, who represent the fastest growing group of cell phone users. A study by the American Academy of Sleep Medicine found that excessive use of cell phones may make teens and young adults restless, more susceptible to stress and fatigue, and more likely to have difficulty falling and staying asleep. Other sociologists have noted that excessive cell phone use by teens may lead to compulsive calling and texting, perhaps even leading to addiction. Accounts of teens' sexting—sending sexual content over a cell phone—and cyberbullying via a cell phone are also increasingly in the news. Using the cell phone while driving—either to send text messages or to make calls—is perhaps the most worrisome effect of the cell phone's intractable place in society, and studies consistently show that young people are far more likely to engage in this risky behavior, when compared with adults.

"Whether cellphones are safe is a matter of personal interpretation of the data, but they are not unsafe enough to motivate most people to stop using them."[16]

— Guy Klemens, author of *The Cellphone: The History and Technology of the Gadget That Changed the World.*

What is clear is that given the ubiquity of cell phones today, research to investigate and understand these and other potential impacts to health and society will continue for years to come. Just as certain is that even in the absence of conclusive evidence, the use of cell phones will continue unabated. As Klemens puts it: "Whether cellphones are safe is a matter of personal interpretation of the data, but they are not unsafe enough to motivate most people to stop using them."[16]

Facts

- According to the American Association of Retired Persons (AARP), 56 percent of cell phone users over the age of 65 cite safety as the primary reason for owning a cell phone.

- The CTIA, the wireless industry's trade association, provides $1 million a year to fund cell phone research. The group has concluded that no health risk is associated with the use of cell phones.

- According to the CTIA, Americans talk on cell phones 2.26 trillion minutes annually.

- Today's iPhone has more processing power than the North American Air Defense Command did in 1965.

- A study by the Institute for Social Research at the University of Michigan revealed that 83 percent of people believe that cell phones make their lives easier.

- According to a 2011 survey, a quarter of all households in the United States do not own a landline telephone and rely solely on cell phones.

Does Cell Phone Use Lead to Cancer?

Cell phone use has been associated with a variety of potential health effects, including headaches, fatigue, and reproductive problems, but the question of whether the RF radiation that is emitted from cell phones causes cancer over time is unquestionably at the center of the debate. Guy Klemens says, "The possibility that the cell phone is a slow killer holds [the most] interest for the public. Much of the miasma of threat arises from the mystery of the invisible. Holding a 0.2 watt light bulb near the body would probably not strike most people as a dangerous or damaging activity. But the electric and magnetic fields surrounding and emanating from cellphones lie outside the visible spectrum. And so people wonder what is coming out of their handsets."[17]

Cell Phones and Cancer

Cancer results when abnormal, or mutated, cells reproduce and spread. A cancerous tumor, for example, starts when a cell's DNA is damaged, which may cause it to reproduce at a high rate, forming a mass of cells called a tumor. The mass can be benign, or harmless, unless one of the cells is mutated. If this cell begins reproducing other mutated cells, they can impede the normal functioning of surrounding cells. When the tumor remains localized, it can often be removed in its entirety. If the mutated cells break free from the tumorous mass and enter the bloodstream or surrounding organs and tissue, however, the cancer can take over vital organs and lead to death.

The specific mechanisms that trigger cancer, and cause its progression, are largely unknown. Scientists have identified certain stimuli that may initiate the cancer process by damaging a cell's DNA. For example, scientists have identified chemicals called polyaromatic hydrocarbons that attach themselves to the DNA of smokers, which may account for the incidence of cancer among smokers, but the exact causes remain elusive. As Klemens puts it:

> Adding to the uneasy sense of the unknown [regarding RF radiation] is the mystery that surrounds tumors and cancer. A tumor does not come with an indicator of its origin, which leaves doctors, patients, and the public searching for possible causes. Even when such a cause seems to present itself, there is still a strong random quality. If smoking caused lung cancer, for example, everyone who smoked would get lung cancer, and yet many do not. Cause and effect is difficult to assign with cancer, so doctors instead deal with risk levels. All activities fall somewhere on the scale of risk, with a low risk indicating safety. But where the safe risk level lies is undefined.[18]

While there is no consensus on whether cell phones increase the risk of developing cancer, one of the largest studies on the effects of cell phone use will undoubtedly stir the current debate for years to come. The report, published in May 2011 by the World Health Organization (WHO), declared that it is "possible" that cell phones could lead to cancer. The statement, issued by the International Agency for Research on Cancer (IARC), an advisory panel to WHO, marks the first time that a major health organization has suggested that cell phones may be carcinogenic.

"The possibility that the cell phone is a slow killer holds [the most] interest for the public. Much of the miasma of threat arises from the mystery of the invisible."[17]

— Guy Klemens, author of *The Cellphone: The History and Technology of the Gadget That Changed the World.*

The WHO Report

The IARC has evaluated close to a thousand suspected carcinogens, assigning them to one of five classification groups. For example, the group has found that 107 are carcinogenic to humans,

Cell Phone Radiation: An Indirect Carcinogen?

Because the nonionizing radiation emitted from cell phones cannot strip electrons away from atoms and damage DNA directly, many scientists believe that this type of radiation cannot directly initiate cancer. According to Siddhartha Mukherjee, physicist and author of the Pulitzer prize–winning book *The Emperor of All Maladies: A Biography of Cancer*, there may be a less direct, and yet unknown, way that cell phones lead to cancer:

> It is possible for something to be a carcinogen without directly damaging DNA. Some chemicals might activate growth pathways or survival pathways in cancer cells (eventually damaging DNA and mutating genes—but indirectly). Exogenous estrogen, for instance, activates growth pathways in breast cells and can cause breast cancer but doesn't damage DNA. Others may provoke inflammation, creating a physiological milieu in the body that allows malignant cells to grow and survive. Yet others—the class of substances that we know least about—might not damage DNA directly but chemically modify genes so that their regulation is changed. These substances are like the dark matter of the carcinogenic world: they are barely visible to our current tests for carcinogens and thus lie at the boundaries of the knowable universe.

Siddhartha Mukherjee, "Do Cellphones Cause Brain Cancer?," *New York Times*, April 13, 2011. www.nytimes.com.

including asbestos and tobacco. Of the 59 that are classified as "probably" carcinogenic is the human papillomavirus, which is associated with cervical cancer in women. Of the 266 factors that are "possibly" carcinogenic are a variety of industrial chemicals, gasoline engine exhaust, lead, coffee, and, most recently, cell phones.

The panel arrived at their decision to classify cell phones as possibly carcinogenic after reviewing dozens of published studies detailing how the type of nonionizing radiation emitted from cell phones affects animals and humans. The body of research suggests an association between cell phones and certain types of tumors, including cancer of the parotid, which is a salivary gland near the ear; acoustic neuroma, a tumor that occurs close to where the ear meets the brain; and glioma, a rare but deadly form of brain tumor.

The largest and longest study of cell phone use and cancer evaluated by the WHO is called Interphone. Organized by the IARC, this vast research effort covered 13 countries, including Canada, Israel, and several Western European countries and over 10,000 subjects aged 30 to 59. The results were published in the *International Journal of Epidemiology* in 2010 and found no overall link between cell phone use and brain cancer; the study suggested that cell phone use actually decreased the risk of glioma by 19 percent. Most scientists believe that this confusing finding reflects a methodological error or bias, as there is no known mechanism whereby cell phones could have this biological effect. At the same time, Interphone showed that heavy cell phone users were 40 percent more likely to develop gliomas.

In another study published in the *American Journal of Epidemiology*, data from Israel revealed a 58 percent higher risk of parotid gland tumors among heavy cell phone users. The study tracked 1,800 Israeli citizens, finding that people who held a mobile phone against their head for several hours a day were more likely to develop this rare form of salivary gland cancer. The study also reported that rural cell phone users were at a higher risk as their cell phones emit higher radiation levels because of their distance from cell towers.

Another report that showed an association between cell phones and cancer came out of Sweden, where scientists analyzed 16 studies in the journal *Occupational and Environmental Medicine*. The body of research showed a doubling of risk for acoustic neuroma and glioma after 10 years of heavy cell phone use.

WHO Report Gets a Mixed Response

Reaction to the WHO report is mixed. Sanjay Gupta, a neurosurgeon and medical correspondent for CNN, echoed the views of others in the scientific community when he declared that the WHO findings "dealt a blow to those who have long said, 'There is no possible mechanism for cell phones to cause cancer.' By classifying cell phones as a possible carcinogen, they also seem to be tacitly admitting a mechanism could exist."[19]

At the same time, the wireless industry and others point out that the studies evaluated by WHO do not provide sufficient data to make a causal association between cell phone use and cancer; rather, the findings highlight the need for more research. As Jonathan Samet, the head of IARC and an epidemiologist at the University of Southern California, remarked: "We found some threads of evidence about how cancer might occur but have to acknowledge gaps and uncertainties."[20] The WHO findings will likely guide future research; at the same time, some feel that the credibility of WHO may make it difficult for critics to dismiss the possibility that cell phone use may lead to cancer.

Other Research

While the debate about cell phone safety is far from resolved, people are incontrovertibly relying more heavily on their cell phones. In addition to the WHO report, scientists have conducted hundreds of studies over the last 15 years to assess the biological effect of this trend. A look at the body of published research reveals conflicting findings, with studies supporting both sides of the issue.

Much of the evidence, however, suggests that RF radiation from cell phones has little biological impact. For example, multiple research projects that have exposed rats to radiation comparable to that emanating from a cell phone have failed to find a consistent

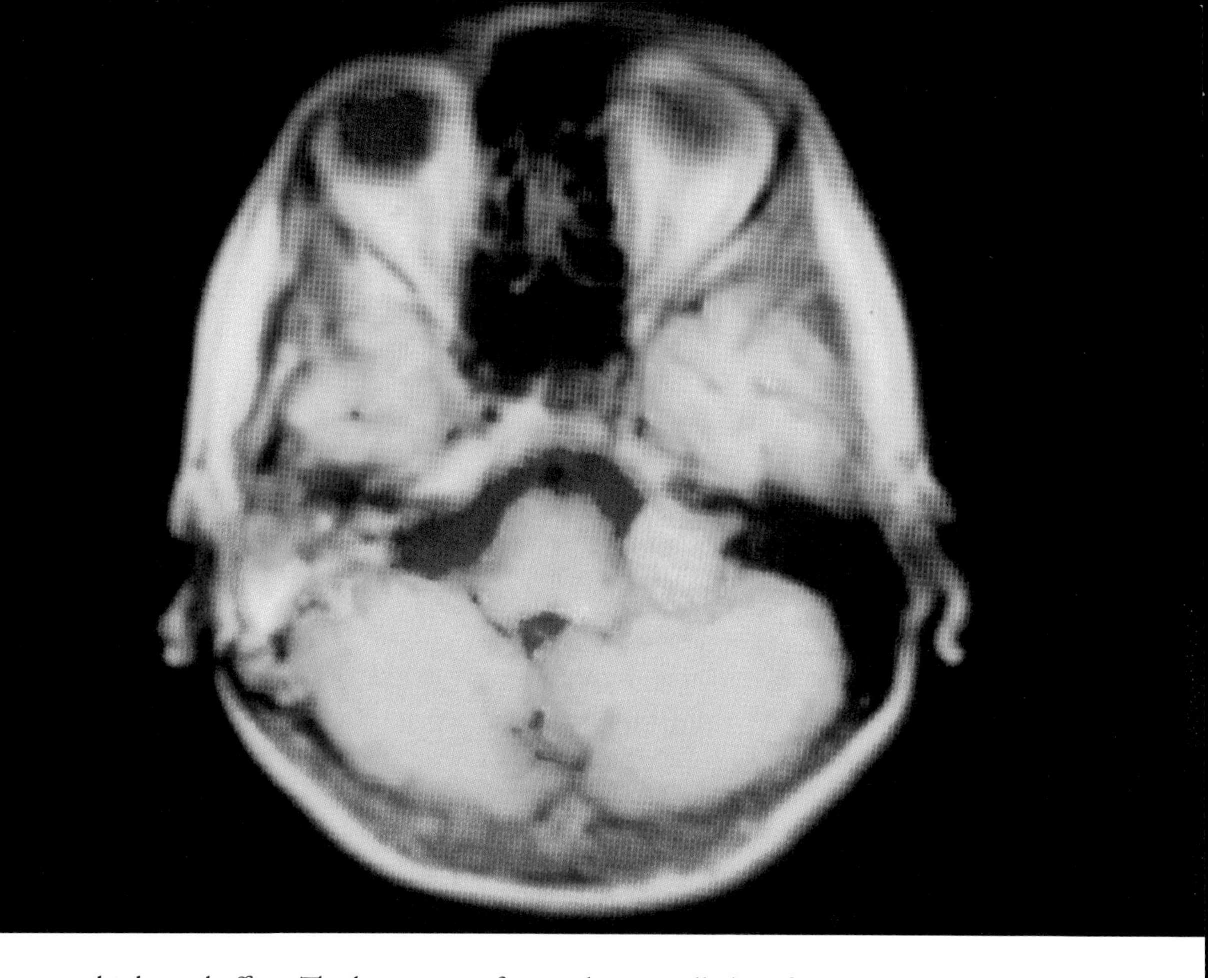

The International Agency for Research on Cancer recently concluded that a link might exist between cell phone use and certain brain tumors. A brain scan shows one such tumor, an acoustic neuroma—the yellow spot on the right between the inner ear and the brain.

biological effect. The largest set of animal tests, called Perform-A, was organized by the European Commission over an eight-year period. The results, released in 2007, found no evidence of a link between cell phones and cancer in the mice and rats studied. Earlier studies reported similar findings: In a 2002 Australian study looking at lymphoma, mice were exposed to radiation comparable to that emitted by cell phones an hour a day, five days a week, for 24 months. Researchers found no increases in lymphoma or other cancers. Two studies in Texas in the late 1990s exposed mice prone to developing breast cancer to extensive radiation for 78 weeks, finding no acceleration in the rate of this type of cancer.

Similarly, many of the epidemiological studies on humans have failed to find a definitive association between cancer and cell phone use. The first large-scale study in Denmark used government cancer records and cell phone records to study whether heavy cell phone

Warning Labels on Cell Phones

Several states have introduced new bills that would require health warning labels on cell phones. In Oregon in 2011, for example, a bill was introduced by Senator Chip Shields that would make warning labels for all new cell phones and cell phone packaging sold in the state mandatory. The bill intends to protect consumers from the possible health effects of cell phones and provide guidelines on how to use them safely.

The city of San Francisco and the states of Pennsylvania and Maine are considering similar mandates. In Maine, for example, state representative Andrea Boland, the first to propose such legislation in 2010, reintroduced the bill in 2011. As currently written, the bill would require the following label to be affixed to all cell phones: "WARNING: Federal health safety standards have yet to be established for nonthermal effects [non-heat-related effects of cellular radiation] of cellular telephone radiation, which have been identified as reasons for health safety concerns, such as brain tumors." Since the science on the health effects of RF radiation from cell phones is inconclusive, these bills are the subject of contentious debate, and to date no legislation has been implemented.

Kent German, "Cities and States Consider Cell Phone Radiation Laws," cnet.com, June 2, 2011. www.cnet.com.

use is predictive of cancer. According to the report's 2001 findings, no association exists.

According to physicist Michael Shermer, RF radiation is simply too weak to break chemical bonds or damage DNA in ways that could generate mutations that could lead to cancer. As Shermer writes in *Scientific American*: "A cell phone generates radiation . . .

that is 480,000 times weaker than UV rays [from the sun]. . . . If the bonds holding the key molecules of life together could be broken at the energy levels of cell phones, there would be no life at all because the various natural sources of energy from the environment would prevent such bonds from ever forming in the first place."[21]

Eric Swanson, professor of nuclear physics at the University of Pittsburgh, also believes that it is virtually impossible for cell phones to cause cancer. In making his case, Swanson compares cell phone radiation with that of visible light:

> Visible light is not dangerous because it does not have enough energy to damage DNA. The amount of light you are exposed to does not make a difference; it is the energy level, not the amount that matters. That's why your skin can happily spend countless hours, day after day, exposed to artificial lighting. What about cell phones? They typically . . . correspond to an energy that is one million times less than visible light. We know that the human body has evolved to not get cancerous when exposed to visible light. What are the odds that it will develop cancer when it is exposed to much less energetic radiation? Not high.[22]

No Dramatic Increase in Brain Cancers

Many scientists maintain, moreover, that if cell phones caused brain tumors or other cancers, a worldwide increase in such cancers would have occurred as more and more people started using these devices. According to the National Cancer Institute, no increase in brain or other nervous system cancers appeared between 1987 and 2005, despite a dramatic increase in cell phone usage during that time period.

The absence of a dramatic increase in brain cancers, coupled with inconclusive evidence of a biological mechanism by which cell phone radiation might cause cancer, leaves many skeptical of an association between cancer and cell phones. As Meir J. Stampfer, a professor at the Harvard School of Public Health, says, "If you look at brain cancer around the world over 25 years that cellphones have been in use, there's no suggestion at all of any increase in rates. In science, unlike math, we can't have absolute certainty, but in the scheme of things, this is not a health risk I would be concerned about at all."[23]

Too Early to Draw Conclusions

With so many conflicting reports, the issue remains murky. As critics point out, some of the studies that found no link between cell phones and cancer were poorly designed and methodologically flawed. For example, the Danish epidemiological study that is widely cited as evidence that cell phone use is safe failed to include children and young adults, who may be especially vulnerable to the effects of RF radiation. At the same time, 92 percent of the subjects tracked had used a cell phone for less than five years; since cancer can take decades to develop, these results may not be predictive of long-term use effects.

"If you look at brain cancer around the world over 25 years that cellphones have been in use, there's no suggestion at all of any increase in rates."[23]

— Meir J. Stampfer, professor at the Harvard School of Public Health.

One of the most outspoken critics of the body of published research is Ronald Herberman, head of the University of Pittsburgh Cancer Institute. Herberman stated in a 2008 congressional hearing that "most studies claiming that there is no link between cell phones and brain tumors are outdated, had methodological concerns, and did not include sufficient numbers of long-term cell phone users to find an effect, since most of these negative studies primarily examined people with only a few years of phone use. . . . In addition, many studies defined regular cell phone use as 'once a week.'"[24]

Herberman went so far as to issue the following memorandum to his faculty and staff in 2008: "Although the evidence is still controversial, I am convinced that there are sufficient data to warrant issuing an advisory to share some precautionary advice on cell phone use."[25] Defending his decision to issue this proclamation, which was highly publicized—and criticized by many in the scientific community—Herberman stated: "We shouldn't wait until definitive information comes out. By then, we might have a virtual epidemic on our hands."[26]

Keith Black, chairman of neurosurgery at Cedars-Sinai Medical Center in Los Angeles, supports this view that it is simply too early to unequivocally declare cell phones harmless: "The biggest problem we have is that we know most environmental factors take several decades of exposure before we really see the consequences."[27] Even Klemens concludes that "the inability of researchers to find a clear mechanism for electric and magnetic field induced diseases through experimentation does not, of course, mean that they do not exist. With mysterious illnesses such as cancer, such a discovery may be a long way in the future, if it even exists."[28]

A Call for More Research

With such a large array of scientific opinion regarding cell phone safety, whether there is a pressing need to implement public health measures to reduce exposure to RF radiation is uncertain. Despite a body of data that suggests no probable cause for concern, no one seems willing to say, incontrovertibly, that cell phones are completely harmless. Rather, the universal agreement among scientists appears to be that more research is needed.

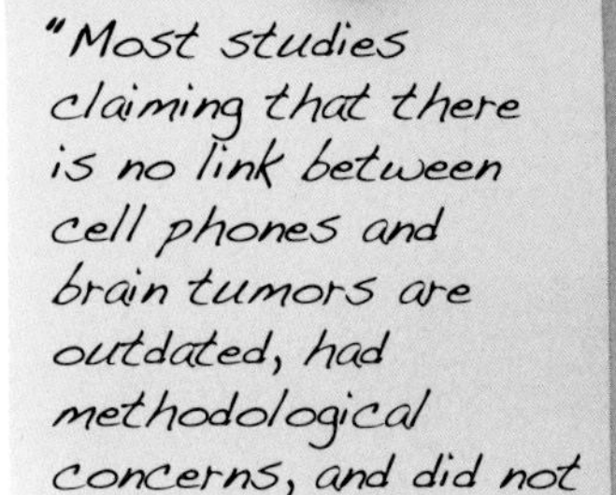

— Ronald Herberman, director of the University of Pittsburgh Cancer Institute.

Until the exact health implications related to long-term cell phone use are established, many individuals and agencies advocate the use of precautionary efforts to reduce exposure to RF radiation. Since exposure falls rapidly with increasing distance, safety measures include simple actions such as wearing a headset or earpiece to limit direct exposure to the brain and refraining from carrying the device close to the body. Other advisements focus on limiting use when the signal is weak since radiation increases as the cell phone searches for a signal.

Devra Davis describes why she believes these measures should be mandatory:

> After ten years of use, increased risks from tobacco and asbestos were not clearly evident, yet nobody today doubts that we waited far too long before addressing these important health hazards. For the sake of our children and

grandchildren, we should promote simple precautions to reduce direct exposure to the brain by using headsets, speaker phones, and texting. This will protect us from whatever health hazards may emerge decades later and also encourage safer development of this revolutionary technology in the meantime.[29]

Governmental Mandates

Several countries outside of the United States have taken action even in the absence of conclusive evidence about cell phone safety. The European Environment Agency, the Finnish Nuclear Regulatory Safety Authority, and the Israeli Health Ministry are among the international groups that have recently declared that the RF radiation from cell phones should be considered a potential health risk. The governments of France, Germany, and India have issued recommendations that exposure should be limited. The French, for example, recently passed legislation that cell phones could not be sold without an earpiece or headset. The French government also prohibits cell phone advertising to children under the age of 14. Similarly, Toronto's Department of Public Health is advising children and teenagers to limit their use of cell phones.

"After ten years of use, increased risks from tobacco and asbestos were not clearly evident, yet nobody today doubts that we waited far too long before addressing these important health hazards."[29]

— Devra Davis, founder of Environmental Health Trust and author of *Disconnect: The Truth About Cell Phone Radiation, What the Industry Has Done to Hide It, and How to Protect Your Family.*

Undeniably, even without hard data about how cellular technology impacts health, cell phones are firmly entrenched in modern life. With nearly every young person in America using a cell phone today, usage is highly unlikely to be curtailed. Klemens notes the psychological impact of this trend:

> As cellphones . . . become a common part of life, their perceived threat may diminish. Unlike the substantial brick-style phones that existed up until the late 1990's, the slighter handsets seem more like small electrical devices, rather than portable radio stations. The cellphone has also lost its status as a new and exotic invention and has found acceptance. As with most novel developments, familiarity dulls the fear of the new.[30]

Many scientific studies have found no adverse health effects from the radiofrequency radiation emitted by cell phones. In a 2005 study in Italy, 1,500 rats living inside habitation units (pictured) were exposed to RF radiation daily. At the end of their natural life spans, autopsies were done to look for possible tumors or diseases.

With more and more people accepting cell phones as an indispensable part of modern life, the scientific community will continue its quest to ascertain if these ubiquitous gadgets are safe. Currently under way is one of the most ambitious studies to date addressing cell phone use and its possible link to brain cancer and other health effects. Launched in Europe in March 2010, the study, known as COSMOS, will track the health of 250,000 cell phone users over the age of 18 for 20 to 30 years. Participants will complete a questionnaire about their lifestyle and cell phone habits. Scientists will supplement this self-reported data with health records and cell phone company records as they search for links between cancer and cell phone use. At the same time, the National Institutes of Environmental Health Sciences (NIEHS) is conducting a laboratory rodent study—the largest such study to date—to assess the effects of radiation exposure similar to that associated with cell phone use. Whether the results from these and other studies will yield more-definitive data on the health effects of cell phone exposure remains to be seen. Multiple generations of lifelong exposure may be required to know for sure whether these effects are harmful.

Facts

- The RF radiation emanating from cell phones is a billionth the intensity of known carcinogenic radiation such as UV radiation or X-rays.

- A cell phone cannot be sold in the United States unless its SAR number, indicating the amount of radiation exposure associated with a particular device, is below 1.6 watts per kilogram. In Europe, the maximum allowable SAR is 2.0 watts per kilogram.

- Both the American Cancer Society and the National Cancer Institute have stated that existing data on a cell phone's impact on health is reassuring.

- Exposure to RF radiation comes from a cell phone's antenna. The amount of exposure dramatically decreases as the phone is moved away from the body.

- Cell phones typically emit the most radiation when they initially establish contact with a cellular tower.

- Professor Lennart Hardell of Sweden found that brain cancer is four to five times more prevalent in young adults who started using cell phones as teenagers than in young adults who did not.

- Whether previous studies would even apply to today's cell phone users is unknown, as people's cell phone habits have changed dramatically in recent years.

Do Cell Phones Cause Other Health Problems?

A possible link between cell phones and cancer is not the only health concern scientists are exploring. A flurry of recent studies suggests that cell phones might adversely impact health in other ways. One of the most common concerns is that the RF energy from cell phones heats biological tissue, which is why some people report that their ear gets warm after they've held a cell phone against it for an extended period of time. The radiation causes molecules in the tissue to vibrate faster, in the same way that a microwave heats food. As Keith Black writes: "What microwave radiation does in most simplistic terms is similar to what happens to food in microwaves, essentially cooking the brain. So in addition to leading to a development of cancer and tumors, there could be a whole host of other effects."[31]

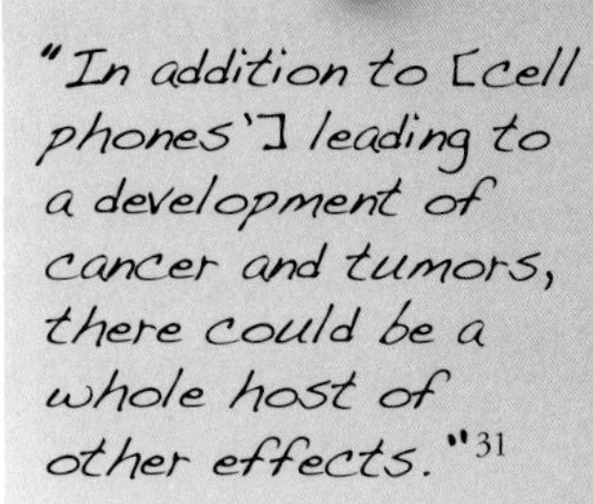

"In addition to [cell phones'] leading to a development of cancer and tumors, there could be a whole host of other effects."[31]

— Keith Black, chairman of neurosurgery at Cedars-Sinai Medical Center.

According to the FDA, an area of the body that appears to be especially vulnerable to tissue heating is the testes. This is because the blood flow to this area is insufficient to help carry away excess heat. A particular concern for men who carry their phones in their pockets or on a belt clip is testicular heating, which could impact male fertility. While no

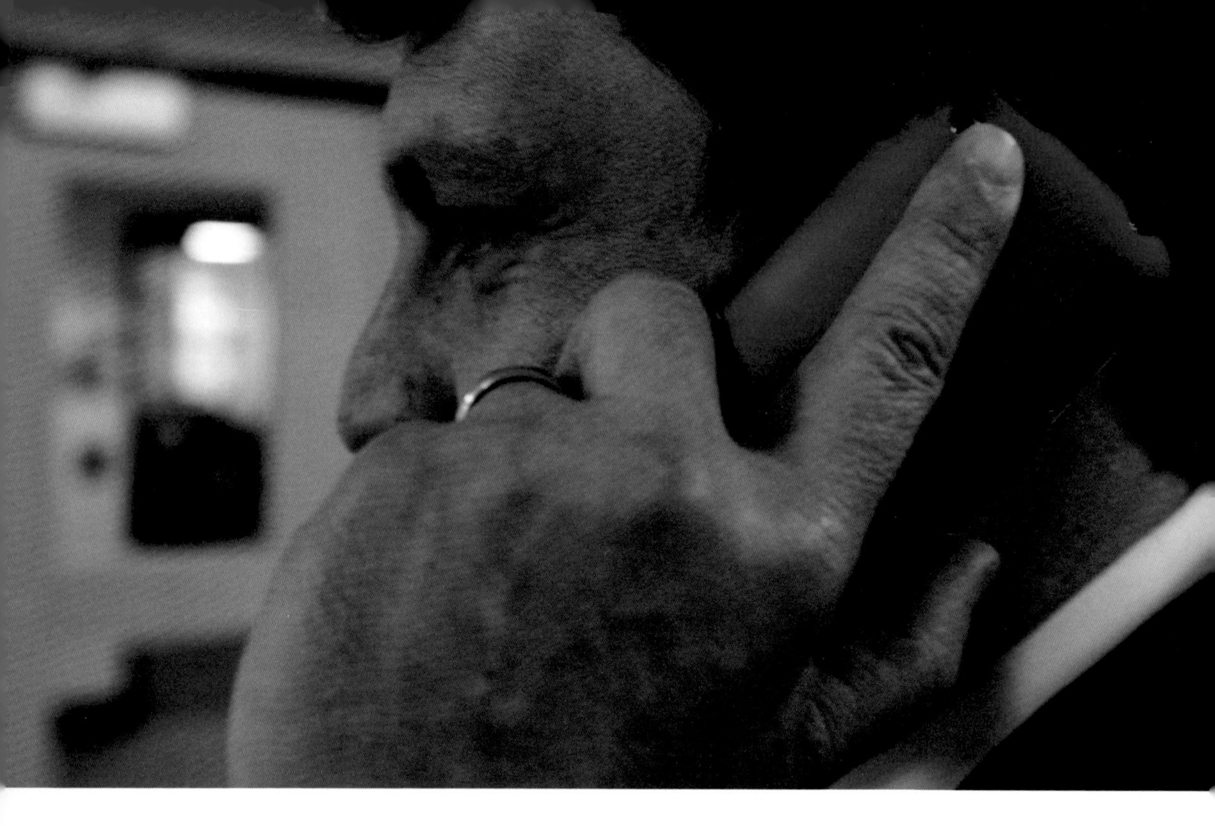

After a long conversation by cell phone, some people report that their ear feels warm. The heating of biological tissue caused by RF energy from the phone is a concern that scientists are researching.

conclusive evidence to prove this claim has been found, some limited data suggests that cell phones, whether through overheating or some other mechanism, can affect the male's capacity to sexually reproduce.

Cell Phone Use and Male Fertility

Cell phone use has soared in the last decade or so. At the same time, male fertility in industrialized countries has decreased. For example, sperm counts among British men have fallen 29 percent over the past fifteen years. That this decrease has occurred over a relatively short period of time suggests that environmental factors may be at play. For example, researchers have speculated that increasing stress, smoking, obesity, and other lifestyle factors have contributed to these plummeting rates. A small but growing number of scientists are starting to believe that the coincidental drop in male fertility and the increase in cell phone use, which has soared in recent years, is no coincidence.

While science has yet to establish that cell phones lead to diminished fertility in men, animal studies suggest that the sperm-making cells in the testes may be particularly vulnerable to the nonionizing radiation or heat that cell phones generate. In one animal study in 2007 at the Medical College of Wisconsin, researchers found that rats exposed to six hours of daily cell phone radiation for 18 weeks had higher rates of sperm cell death than rats that were not exposed. The authors of the study suggested that carrying cell phones near reproductive organs could negatively impact fertility. Likewise, a 2009 rat study at the Melaka Manipal Medical College in India found that the radiation from cell phones negatively affects semen quality and could impair male fertility. Other animal studies have reached far different conclusions, however: A 2005 study in Brazil, for example, found that radiation emitted from cell phones does not impair testicular function in rats in any way.

While these conflicting studies leave the issue unresolved, healthy sperm are indisputably crucial to human reproduction. As Devra Davis comments: "Sperm are among the fastest and speediest-growing cells in the body. They must travel heroic distances to succeed in fertilizing an egg. . . . Sperm have a head and tail that propel them forward. . . . Sperm that swim fastest and straightest have the best chance of success and tend to be those with the longest, strongest tails to whip them along."[32]

Sperm Damage

Professor Ashok Agarwal, director of the Andrology Laboratory and Reproductive Tissue Bank at the Cleveland Clinic, is among those who believe that heating or some other mechanism related to cell phones may be causing sperm damage that has led to the falling fertility rates in Britain and other parts of the world. In a 2007 study led by Agarwal, researchers compared the sperm health of 361 men, who were divided into three test groups according to their cell phone habits: those who never used a cell phone, those who used a cell phone less than two hours per day, and those who used a cell phone at least four hours per day. Agarwal found that the group of men who used their phones the most—that is, at least four hours per day—had a 25 percent decrease in their sperm count compared with the group of men who never used a cell phone. Equally alarming, the men who used their cell phones the most displayed diminished sperm quality, measured by such attributes as the swimming ability of sperm, which is crucial to conception. When researchers looked at the appearance of the sperm under the microscope, those from men who used cell phones the most fared poorly: The heaviest cell phone users had a 50 percent drop in the number of properly formed sperm.

While these results appear compelling, the study relied on the subjects' self-reported cell phone habits, which may not have been accurate. It also did not account for other lifestyle factors that could contribute to lower sperm counts.

To account for these variables, Agarwal devised a more controlled laboratory experiment to assess what happens to sperm samples when they are exposed to radiation similar to that emanating from a cell phone. Researchers exposed sperm samples to cell phone radiation or no radiation at all. The team found that 85 percent more free radicals—unstable molecules in the body that can damage healthy tissue—were generated by the sperm samples exposed to cell phone radiation when compared with sperm samples that had not been exposed. Agarwal speculates that cell phone radiation creates a state of oxidative stress, which negatively impacts the body's cells and tissues, including sperm health and semen quality: "The results of our study were significant and striking. The data lead us to speculate that carrying a cell phone in a pocket in talk mode leads to deterioration of sperm quality through oxidative stress."[33]

"The results of our study are significant and striking. The data lead us to speculate that carrying a cell phone in a pocket in talk mode leads to the deterioration of sperm quality."[33]

— Ashok Agarwal, director of the Andrology Laboratory and Reproductive Tissue Band at the Cleveland Clinic.

At the same time, Agarwal acknowledges that more study is needed, stating that "one of the main differences between our experimental conditions and real life is the multiple tissue layers that separate the cell phone and the reproductive organs in vivo [in a living organism]. Further studies are needed to allow valid extrapolation of the effects seen under in vitro [in an artificial environment] conditions to real-life conditions, and these are already under way in our laboratory."[34]

Can Cell Phones Help Fight Alzheimer's Disease?

After years of speculation that the RF radiation from cell phones may cause cancer and other ailments, a new study suggests that cell phones might actually fight Alzheimer's disease. An irreversible, progressive brain disease, Alzheimer's destroys many types of mental function, including memory and cognitive skills. It is most often diagnosed in people over 65 years of age. Although the cause of Alzheimer's disease is not fully known, the disease appears to be associated with sticky brain deposits known as beta amyloid plaques, which can build up between nerve cells and impair their function.

In the study, published in the *Journal of Alzheimer's Disease* in 2010, scientists examined the effects of cell phone radiation on mice that had been genetically modified to develop beta amyloid plaques. These mice typically developed the first Alzheimer's symptoms at around 6 months. The genetically modified mice and a control group were exposed to two hours of cell phone radiation daily for 7 to 9 months. Researchers found that if the exposure began before the first symptoms appeared, the Alzheimer's-prone mice were less likely to develop cognitive deficits. Those that were exposed after the onset of symptoms saw improvements in memory function after several months of exposure. While these findings are striking, many scientists caution that it is too early to know whether they have any relevance to humans.

Possible Effects on Reproductive Hormones

Another study by research teams at Queen's University in Canada and the Medical University of Graz in Austria has identified yet another way that cell phones could affect male fertility. The study, published in 2011, found that the radiation emitted from cell phones may have an effect on the male reproductive hormones that are produced in the brain. Specifically, researchers found that men who used cell phones had higher levels of testosterone circulating in the body but they also had lower levels of luteinizing hormone (LH), a reproductive hormone that is secreted by the pituitary gland in the brain. LH is crucial to reproduction as it converts the basic circulating type of testosterone into the more potent form that is associated with fertility.

As lead researcher Rany Shamloul says: "Our findings were a little bit puzzling. . . . The results we [found] suggest that there could be some intriguing mechanisms at work."[35] While this study raises interesting questions about a cell phone's effect on male fertility, Shamloul concludes that more research is needed to understand the complex interplay between cell phone radiation and male reproductive hormones. As research continues, scientists will continue to tease out the various mechanisms whereby cell phone radiation might have an effect not only on reproduction but also other body processes, including how the brain protects itself.

The Blood-Brain Barrier

Human brains are protected by a bony skull. Another form of defense—an internal mechanism—was discovered over a century ago. Scientists noticed that if blue dye was injected into an animal's bloodstream, tissues throughout the entire body would turn blue, except for the brain and spinal cord; researchers speculated that some sort of protective mechanism existed to protect the brain from taking in the wrong substances. Scientists now know that the blood-brain barrier (BBB) restricts the passage of bacteria or other harmful matter into the brain.

The BBB is made up of endothelial cells that fit so tightly together that most substances cannot pass through them. Some materials, such as glucose when it is attached to specific proteins, are able to cross the semipermeable BBB. For the most part, however, only small molecules soluble in fat clear the barrier. This system operates quite effectively to protect the brain from bacteria and other foreign agents that could cause harm.

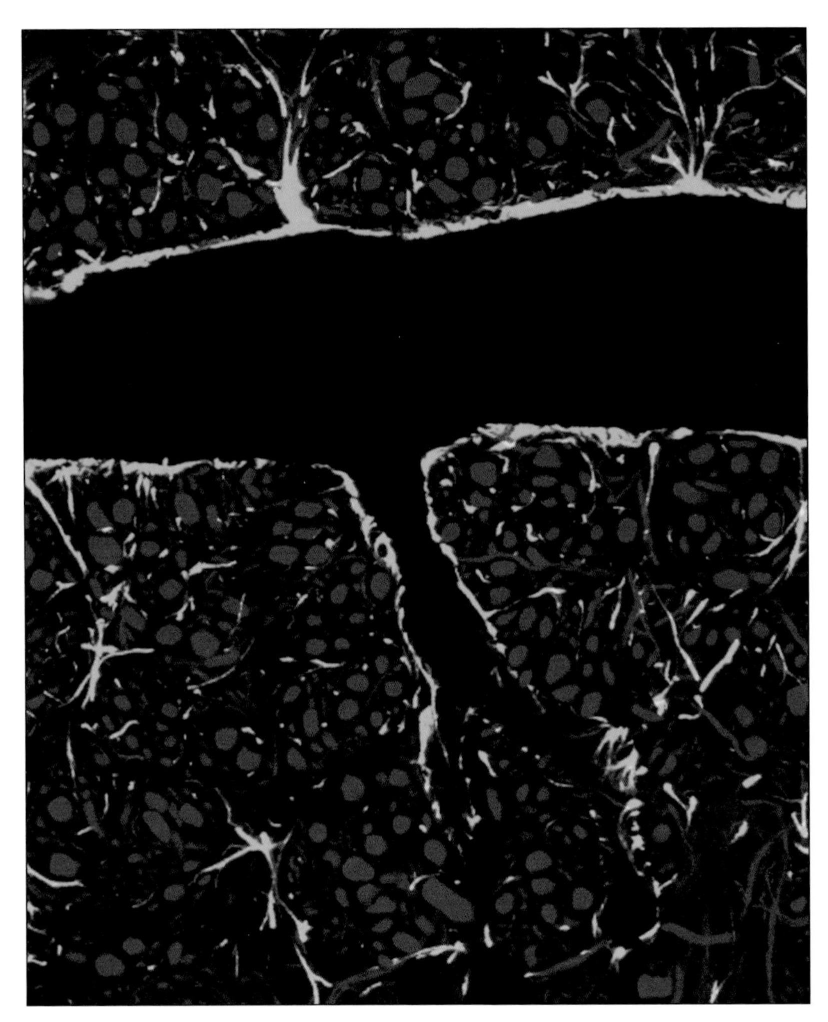

Swedish scientists, searching for a more direct means of treating brain tumors, have found that RF radiation can open the blood-brain barrier that protects the human brain from bacteria and other harmful agents. This finding offers promise for the treatment of brain cancer but raises concerns about the effects of RF radiation on healthy brains. A light micrograph reveals a section through a blood vessel in the brain, showing the arrangement of cells that form the blood-brain barrier.

Certain factors can break down the BBB, however. For example, high blood pressure and certain infectious agents can open up the BBB, as can injury to the brain that causes inflammation or pressure. Once this vital membrane is breached, the brain will more readily absorb anything circulating in the blood—toxic chemicals or infectious agents that would not normally enter the brain, for example.

Leakage

Over three decades ago, a neuroscientist named Allan Frey discovered that RF radiation could also relax the BBB. In a series of experiments to assess how microwaves at low power affect biological

Do Cell Phones Lead to Insomnia, Headache, and Depression?

New research has linked the RF radiation emanating from cell phones to headaches, sleep disturbances, confusion, and other symptoms. The joint study was conducted by scientists from the Karolinska Institute and Uppsala University in Sweden and from Wayne State University in Michigan. Scientists studied 35 men and 35 women who were exposed either to real radiation equivalent to that emitted by a cell phone or sham radiation—that is, no exposure at all; the subjects were not told whether they had been exposed to real or sham radiation.

Researchers found that exposure to RF radiation from a cell phone before bed disrupts the body's ability to enter the deep stages of sleep, which scientists believe is crucial to human health as it allows the brain and body to recover from daily stresses. Subjects were also likely to report experiencing headaches and difficulties in concentration after being exposed to real radiation versus sham radiation. As lead researcher Bengt Arnetz said, "If you have trouble sleeping, you should think about not talking on a mobile phone right before you go to bed. The study strongly suggests that mobile phone use is associated with specific changes in the areas of the brain responsible for activating and coordinating the stress system."

Wayne State University, "Wayne State University Researcher Gains International Attention for Cell Phone Study," January 22, 2008. www.media.wayne.edu.

systems, Frey injected a fluorescent dye into the circulatory system of rats before exposing them to the radiation. Within minutes of exposure, the dye had leached into the rats' brains. Frey discovered that pulsed microwave radiation similar to that from the cell phones of today produced the greatest biological effects. Frey published his results—that microwaves pulsed at specific modulations could cause leakage in the BBB—in the *Annals of the New York Academy of Sciences* in 1975.

Because a later study failed to replicate Frey's findings, research on the effect of microwaves on the BBB largely halted in the United States. Research is ongoing in many countries outside the United States, however. In Sweden, for example, neurosurgeon Leif Salford wondered if RF exposure could open the BBB so that he could push chemotherapy into the brains of cancer patients. As Salford says, "I knew that the brain usually resists taking anything into it. From where I stood as a neurosurgeon treating cancer, all I wanted to do was to crack into the brain chemically so that we could deliver some of the agents that we knew would kill brain tumors."[36]

In the early 1990s, Salford and his colleagues at the Rausing Laboratory for Experimental Neurosurgery and Radiation Physics did in fact succeed in using radiofrequency radiation to open the BBB of cancer patients so that drugs could be introduced into the brain to treat the cancer. This led Salford to question how this exposure affects the brains of healthy people who are regular cell phone users.

"We now see that things happen to the brains of lab animals after cell phone radiation. The next step is to try to understand why this happens."[37]

— Henrietta Nittby, researcher at the Rausing Laboratory for Experimental Neurosurgery and Radiation Physics.

As they set out to answer this question, Salford and his team confirmed Frey's findings: They found that rodents exposed to as little as two hours of pulsed cell phone signals similar to 2G phones operating at 900 MHz or 3G phones operating at 1900 MHz readily absorbed dyes into their brains. They also found the rodents exposed to these relatively short cell phone exposures displayed significant deficits in learning and were unable to complete tasks that they usually performed with ease, such as making their way out of a simple maze.

Other research out of Sweden by Salford and his colleague Henrietta Nittby found that albumin, an important protein in the blood, leaks into the brain tissue of rats exposed to pulsed cell phone signals. Today, Nittby is involved in potentially groundbreaking research that will assess still other molecular or genetic changes in rodent brains after exposure to cell phone radiation.

As Nittby says, "We now see that things happen to the brains of lab animals after cell phone radiation. The next step is to try to understand why this happens."[37] At the same time that these lines of scientific inquiry remain open in Sweden and other countries, provocative new research from the United States has found a significant association between cell phone use and activity in the human brain.

Cell Phones and Brain Activity

Glucose, a type of sugar, fuels activity in the brain. When the brain is activated—for example, when a person processes visual or auditory signals—the brain cells begin to metabolize glucose at a higher rate. Using brain scans on human subjects, scientists can peer inside the human brain and observe the ebb and flow of glucose during a variety of activities, including cell phone use.

"Cellphones are fantastic and have done much to increase productivity. I'd never tell people to stop using them entirely."[39]

— Nora Volkow, neuroscientist and director of the National Institute on Drug Abuse (NIDA).

Nora Volkow, one of the world's leading neuroscientists and director of the National Institute on Drug Abuse, recently published a groundbreaking study measuring these effects. Her year-long study, published in the *Journal of the American Medical Association* (JAMA) in 2011, reported striking patterns between brain-glucose metabolism and cell phone exposure. As part of her research, Volkow and her colleagues took brain scans of 47 volunteers after they had held a cell phone to their ear for 50 minutes. The phone was active but silent so that she could eliminate the possibility that the changes in brain activity were the result of sound or conversation. When she compared changes in glucose flow, she found the subjects who had their phone on had higher levels of glucose metabolism—a measure of brain activity—even though they were not speaking or listening.

Specifically, two portions of the brain closest to the cell phone antenna—the orbitofrontal cortex and parts of the temporal lobe—showed conspicuous increases in glucose metabolism. The orbitofrontal cortex is broadly associated with emotion, memory, aggression, eating, and a variety of other behaviors. The temporal lobe is critical to language and memory. This leads many scientists

Researchers in Sweden have found that albumin, an important protein in the blood, leaks into the brain tissue of rats exposed to pulsed cell phone signals. A colored scanning electron micrograph shows albumin crystals from a blood clot.

to question whether the cell phone has the capacity to affect the ability to use language or to learn. As Black puts it: "This study raises a lot of questions. Will cellphones impact how we remember things, is there any relation to the risk of Alzheimer's? Will it affect our cognitive ability to manipulate language functions?"[38]

Although Volkow's data are preliminary and the answers to these questions are unknown, the *JAMA* study challenges the longstanding belief that the radiation emitted from cell phones is too weak to have an effect on the brain or other human tissues. Rather, Volkow's findings prove, irrevocably, that cell phones do have a biological effect on the brain, although many scientists are quick to point out that increases in glucose metabolism can occur naturally, such as when a person is thinking, and do not necessarily indicate that cell phones are dangerous.

In light of her findings, Volkow now advises users to keep their cell phones away from their body whenever possible. She has stated that even she uses an earpiece or speaker phone to reduce her own radiation exposure. Despite her precautions, Volkow says that "cellphones are fantastic and have done much to increase productivity. I'd never tell people to stop using them entirely."[39] Volkow's research—and public statements about her own cell phone habits—will surely generate a new line of inquiry into whether these cell phone–induced changes in brain activity are consequential in relation to human health and behavior.

Today, few disagree that more research is needed to determine whether cell phones pose a risk to human health. The final verdict could be years away. Until then, the scientific debate surrounding cell phones will remain heated and unresolved.

Facts

- **Cell phones can trigger allergies: Contact dermatitis, or "cell phone rash," appears in users who are allergic to the nickel used in cell phone fabrication.**

- **In a study of more than 420,000 Danish adults, long-term mobile phone users were 10 to 20 percent more likely to be admitted to the hospital for vertigo or migraine, compared with those who used cell phones for less time.**

- **In one 2003 Swedish experiment, Swedish neurosurgeon Leif Salford found that microwave exposure stimulated neurons associated with Alzheimer's disease in rats.**

- **A study published in the journal *Bioelectromagnetics* suggested that the RF radiation from cell phones created tiny air pockets in the lens of the eye that could cause eye damage or cataracts.**

- **A team of Israeli researchers found a possible link between prolonged exposure to microwave radiation similar to that emanating from cell phones and the development of cataracts in the eye.**

- **A study at the Institute of Biochemistry and Biophysics in Tehran found that microwave radiation reduced the ability of blood to carry oxygen.**

CHAPTER FOUR

How Has Cell Phone Use Affected Driving?

In 2009 Chance Wilcox, 25, was killed because Jeri Montgomery, using a cell phone, made an illegal turn on Interstate 45 in Texas and hit Wilcox's truck. The truck flipped four times and Wilcox was killed instantly. Montgomery was found guilty of negligent homicide. Wilcox's mother said, "I no longer have a son, my daughter no longer has a brother, my family has fallen apart."[40] Unfortunately, many other families can tell a similar painful story.

Recent studies have shown that the phenomenon known as "distracted driving," in which a driver performs an activity such as eating, using a compact disc player, or ducking to retrieve an item on the car floor, is increasing dramatically with cell phone use. The number of distractions such as texting, using a handheld phone, or using a hands-free phone device has meant that drivers have newer distractions. While most law enforcement officials argue that any distraction is a bad distraction, many others contend that cell phone use cannot be stopped, and drivers must learn, instead, to incorporate cell phone use while driving in a safe manner.

New Technology

A cell phone is a relatively new technology that has only been adopted recently. According to the CTIA, more than 114 million people now subscribe to wireless service—up from 7.5 million in 1991. Cell phone ownership increased dramatically in just a little over 10 years. In 1999 only 33 percent of Americans had a cell phone. By 2008, 91 percent had a cell phone. Texting is even more recent. The average monthly volume of text messages was 1 million in 2002. By 2008 it was 110 million.

Although the mobile phone has been around since the 1980s, it was primarily used in businesses as a way of communicating between employees in a variety of locations, such as construction workers who could communicate from a job to the home office. Since that time, cell phones have become smaller and cheaper as communications companies rushed to make them more convenient and easier to use and expanded their range by building signal towers as quickly as possible. Competition between cell phone companies has resulted in phones so easy to use that they are now indispensable for personal use. Today, many Americans forgo a traditional land line and only have a cell phone. Their portability allows cell phones to be used anywhere, including while driving. Yet, because the cell phone is such a new technology, society has been playing catch-up in studying its impact on accident rates and whether laws should be passed to curb its use while driving.

Statistical Evidence

Statistics seem to confirm the dangers of driving while using a cell phone. According to the National Highway and Traffic Safety Association, of those killed in distracted-driving-related crashes in 2009, 995 involved reports of a cell phone as a distraction—and 18 percent of the fatal crashes involved cell phones.

The statistics are even direr for teens. In 2009, 21 percent of fatal car crashes involving teenagers between the ages of 16 and 19 occurred due to cell phone use. Using statistical analysis, experts say that this figure is likely to grow by 4 percent each year. Teen drivers are 4 times more likely than adults to be involved in a car crash or near crash because of talking or texting on a cell phone. In response, 30 states have specific laws that prohibit teens' use of cell phones while driving.

Teens on Texting and Driving

According to the 2010 Pew Research study, cell phones have led to an increase in distracted driving by teens. A particularly disturbing finding is that most of the teens interviewed were unaware of the large amount of research confirming the dangers of texting and/or talking on a cell phone when driving.

Many teens used tactics to avoid being caught by police while texting and driving. One high school boy said, "I wear sunglasses so the cops don't see [my eyes looking down]." Another high school boy said that certain activities were safer, that "just reading a text isn't that bad, it's just reading and then moving on. If you're texting, it's going to take more time when you're supposed to be driving, and that's when most people get in accidents." The findings suggested that just like their adult counterparts, teens were going to use cell phones while driving regardless of the laws.

Quoted in Pew Research Center, "Teens and Mobile Phones," April 20, 2010. www.pewinternet.org.

Concern about the potential dangers of driving while using a cell phone has spawned a number of studies. One study conducted in 2006 by the University of Utah in Salt Lake City compared cell phone use while driving with driving under the influence of alcohol. If the effect on driving was equivalent, researchers thought, then states should pass laws restricting cell phone use in the same way that they have laws and structured penalties for driving under the influence. The study used volunteers, half of whom consumed enough alcohol to be above the legal limit for driving (.08 percent), while other volunteers used some type of cell phone device. All drivers drove in a car simulator that reproduced the scenario of following behind another vehicle that braked suddenly.

The study garnered several interesting results. One was that the researchers found little to no difference in driving ability between those who used a handheld cell phone and those who used a hands-free device. In both scenarios, the participants were equally distracted and made an equal number of errors while driving. Since many states allow drivers to use hands-free cell phone devices with the understanding that they are safer, this result was particularly damning. The study suggested that the reason for more driving errors is not that the driver does not have two hands on the wheel but rather that the driver is distracted by conversation.

When the alcohol-impaired drivers and the cell phone users were compared, several differences were noticeable. The cell phone users had a delayed response time to the braking scenarios and were involved in more accidents than the drivers who drove while legally drunk. The drunk drivers drove closer to the car in front of them and braked more aggressively, but were involved in fewer accidents. The study concluded that "When driving conditions and time on task are controlled for, the impairments associated with using a cell phone while driving can be as profound as those associated with driving with a blood alcohol level over 0.08 percent."[41]

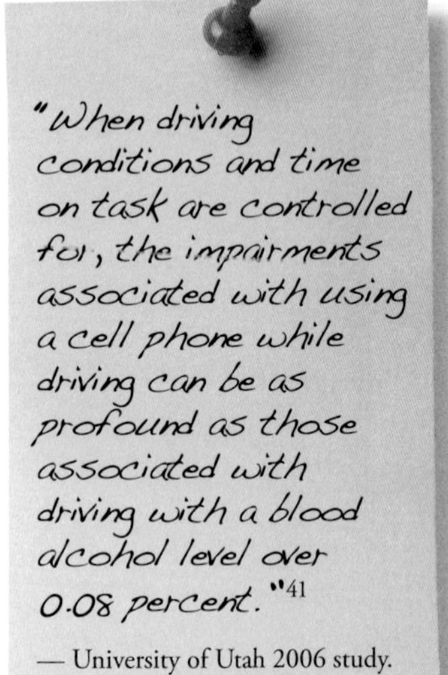

Yet not all studies agree that all cell phone use is dangerous and that zero tolerance is necessary. In 2009 scientists at the Virginia Tech Transportation Institute used real-life scenarios by placing a camera in the cars of volunteers for a year. When footage that involved crashes or near crashes was examined, researchers noted a marked difference in handheld versus hands-free cell phone use. This study concluded that while manipulating or using a handheld cell phone caused more accidents, simply engaging in a phone conversation using a hands-free device did not.

The Brain on Technology

Part of the reason cell phones may be so deadly in the hand or ear of a driver has to do with how the human brain works. When a person directs his or her attention toward sound, the brain's visual capacity decreases. Steven Yantis, a professor of psychological and brain sciences at Johns Hopkins University, says that the effect is as if drivers are seeing the image in their head of the person they are talking to, thereby decreasing their ability to see what is actually in front of them. "When people are listening to a cell-phone conversation, they're slower to respond to things they're looking at. It requires you to select one thing at the cost of being less able to respond to other things."[42]

A 2006 study from neuroscientists Paul E. Dux and René Marois of Vanderbilt University in Nashville, Tennessee, also suggests the brain may be unable to reconcile the input of a cell phone with the task of driving. Marois, associate professor of psychology at Vanderbilt, concluded: "While we are driving, we are bombarded with visual information. We might also be talking to passengers or talking on the phone. Our new research offers neurological evidence that the brain cannot effectively do two things at once. People think if they are using a headset with their cell phone while driving they are safe, but they're not because they are still doing two cognitively demanding tasks at once."[43]

Using a car simulator, an Alabama high school student tries to keep an eye on the road while driving and texting. Studies using simulators such as this one have confirmed the dangers of driving while texting.

"People think if they are using a headset with their cell phone while driving they are safe, but they're not because they are still doing two cognitively demanding tasks at once."[43]

— René Marois, associate professor of psychology at Vanderbilt University.

The researchers used functional magnetic resonance imaging (fMRI) on subjects while they performed two tasks: pressing the appropriate computer key in response to hearing one of eight possible sounds and uttering an appropriate syllable in response to seeing one of eight possible images. The lateral frontal, prefrontal, and superior frontal cortex are involved in cognitive ability. These areas were unable to process the two tasks at once. Rather, the brain resorted to "queuing" the tasks, postponing the second task until it completed the first. Though the brain is able to shift between tasks extremely quickly, the researchers did conclude that driving while using a cell phone would result in impaired driving. Thus, while most people believe that they can multitask, or perform several tasks at once, what they are actually doing is paying continuous, partial attention to several tasks.

Texting

Many cell phone users are also texting while driving. A 2010 study published in the *Journal of Public Health* concluded that texting while driving accounted for 16,141 deaths between 2002 and 2007. Studies have concluded that when a driver texts, reaction time is decreased by 35 percent, steering capability goes down by 91 percent, and a driver is 23 times more likely to be involved in a car accident.

Research using participants in a driving simulator also confirms the dangers of texting while driving. In a study conducted by Frank Drews at the University of Utah, 40 participants were tested in a driving simulator—driving without texting and then driving while texting. Results revealed that participants who were texting and driving at the same time were slower to respond to brake lights in front of them, had less control of steering, and were involved in more accidents.

Across the country, fatalities involving texting while driving are almost a daily occurrence. In 2011 in a suburb of Boston, Massachusetts, just one of hundreds of such fatalities occurred. Thirty-one-year-old Craig Bigos killed a 13-year-old boy on a bicycle while driving and typing a text message on his cell phone. The man lost control of his sport utility vehicle and hit and killed the boy. Stories like this one are repeated daily throughout the country, spawning stricter laws, safety education classes, and other responses.

Texting and Driving While on the Job

One of the areas of biggest concern for employers and the public in general is the use of cell phones by those who drive while on the job, especially long-distance truck drivers and those who drive public transportation vehicles. These drivers have the potential to kill many more people if they drive while distracted.

Using Technology to Curb Cell Phone Use While Driving

Almost all cell phones carry an imbedded GPS device. The FCC mandated that cell phones carry this technology so that 911 operators know where drivers are located when they make an emergency phone call. The device allows emergency operators to not only know where a driver is located but also the driver's latitude, longitude, and speed.

Some manufacturers are building cell phone devices using this technology to prevent drivers from being able to use texting and/or calling features on their phones after a car starts. When the phone detects that the driver is in motion using the GPS function, the phone will no longer operate.

One such device, called Key2SafeDriving, is able to be programmed by a parent and/or employer to enforce preestablished rules. Key2SafeDriving also has a feature that automatically reports attempts to circumvent the device. For example, if a person tries to send a text or make a call, attempts to tamper with the product, or dials 911, the device immediately sends a text message alert.

One particularly dangerous incident occurred in Los Angeles in 2008 when Robert Martin Sanchez, 46, driving a Metrolink commuter train, ran a red light and plowed into a Union Pacific rail car in Chatsworth, California. The crash resulted in the deaths of 25 people and injured 135. Sanchez also died in the crash. The investigation following the crash revealed that the driver had been receiving and sending text messages while driving the train.

A similar incident occurred in Boston in 2009, when a 24-year-old conductor on a metro trolley rear-ended another trolley, sending scores of people to the hospital. The driver admitted that he had been texting his girlfriend. Fortunately, no one died in the incident.

While all drivers are subject to state and federal laws, additional laws have been passed regarding public transportation operators and texting. The US Department of Transportation prohibits drivers of interstate buses and trucks over 10,000 pounds from sending text messages on handheld devices. All federal employees are banned from texting while driving. In addition, many private transportation companies also impose limits. Even when a company has a specific policy prohibiting cell phone use, the policy does not protect the company from being sued. More people who are injured in a crash with someone operating a vehicle for business are suing the employee's company for damages.

In one incident in 2003, the state of Hawaii agreed to pay $1.5 million to a pedestrian struck by a car driven by a state employee talking on a cell phone. The pedestrian sustained permanent brain injuries. In another incident in March 2000 a law associate in Virginia struck and killed a 15-year-old girl while conducting business on her cell phone. The girl's family sued the driver and her law firm for $30 million. Cases such as these have caused employers to enact strict penalties or immediate dismissal of employees caught texting while operating a company vehicle.

Curbing Cell Phone Use

State by state, governments are passing laws that impose a mix of restrictions on cell phone use while driving. No state bans all cell phone use, but some states prohibit cell phone use by certain drivers—30 states ban all cell phone use by novice drivers, while

19 states and DC prohibit school bus drivers from using a cell phone while carrying passengers. Nine states, DC, and the Virgin Islands prohibit all drivers from using handheld cell phones while driving.

Outright bans on handheld cell phones may be counterproductive. According to the Highway Loss Data Institute, an organization funded by insurance firms, no differences are found in collision rates between those using a handheld phone versus those using a hands-free device. According to the US National Highway Traffic Safety Administration, in fact, allowing Bluetooth activity only seems to increase distracted driving. Drivers are under the mistaken impression that using a Bluetooth device is safer and so tend to talk longer and more frequently. Thus handheld bans, ironically, lead to an increase in overall cell phone use, which may lead to even more accidents.

Emergency personnel respond to a head-on crash between a Metrolink commuter train and a Union Pacific freight train in 2008 in Chatsworth, California. An investigation revealed that the Metrolink driver had been sending and receiving text messages while driving the train.

Texting has been banned in 34 states, DC, and Guam. Some organizations, including the Governors' Highway Safety Administration (GHSA), advocate that texting should be banned completely and cell phone use banned for novice drivers. Some states, such as California and New York, prohibit use of a handheld cell phone. Texting bans may also result in more accidents. Many drivers refuse to stop texting but do so more covertly by placing the cell phone in their laps to text. This increases the amount of time it takes to text, and so increases the time the driver's eyes leave the road.

Educational Programs

Some states focus on educational programs, similar to "Click It or Ticket," the program to increase seat belt use. A pilot program in Hartford, Connecticut, and Syracuse, New York, combined increased police enforcement along with education and media coverage to reduce the use of cell phones and texting while driving. The program's slogan was "Cell phone in one hand, ticket in the other." Results suggested that the program increased driver awareness of the current cell phone laws in these states. The program ran for one year starting in 2010 and reduced cell phone use and texting while driving in Syracuse by 32 percent and in Hartford by 72 percent. Laws differed slightly in the two states, which could partially explain the discrepancy. In Syracuse, police officers can stop a driver for cell phone use or texting alone, whereas in Hartford the driver must be committing another offense in order to receive a ticket, such as swerving or driving erratically.

Despite the new laws, a 2010 study conducted by the Highway Loss Data Institute has shown that such enforcement has not changed behavior nor lowered the incidence of crashes related to texting and talking on the phone. This study's surprising results suggested that either more enforcement is needed or that drivers will not change their behavior even if they know they will be ticketed.

Drivers Will Do It Anyway

Another survey by the Insurance Research Council found that 84 percent of cell phone users already believe that using a phone while driving increases the risk of an accident. But 61 percent of those surveyed say they continue to use their phone while driving; 50 percent of drivers between the ages of 18 and 24 do so.

Given the fact that drivers will do it anyway, many legislators think that people should be given guidelines for safe use, just like they are given guidelines about how to eat food safely while driving, and other distracting behaviors. The National Safety Council (NSC) proposes urging drivers to use cell phones only when absolutely necessary, such as to alert the police to a dangerous driving situation or to a drunk driver. According to the NSC, if and when drivers use a cell phone they should plan it in their heads before making the call, to keep it short and precise. In addition, the NSC recommends that the driver inform the person on the other end of the line that he or she is driving and must keep it short.

Other proposed solutions include manipulating the cell phone itself so that it cannot be used while driving. One such technology is called Key2SafeDriving. When the car starts, the application automatically places the driver's cell phone into Safe Driving Mode. Like airplane mode, which is currently available on many cell phones, safe driving mode does not allow the driver to send or receive calls or text messages. Incoming calls are sent directly to voice mail, and incoming text messages are automatically responded to with a message that informs the sender that the person is driving and will respond when he or she arrives at a safe location. As the CEO of Safe Driving Systems, Mike Fahnert, the maker of the product claims, "Our number one goal is to help save lives by empowering parents, business owners, and fleet managers with tools to eliminate mobile phones as a driving distraction and giving them peace-of-mind when their teenagers and employees are behind the wheel."[44]

"Our number one goal is to help save lives by empowering parents, business owners, and fleet managers with tools to eliminate mobile phones as a driving distraction and giving them peace-of-mind when their teenagers and employees are behind the wheel."[44]

— Mike Fahnert, CEO of Safe Driving Systems.

Though use of a cell phone while driving clearly compromises attentiveness and safety on the road, cell phone bans are controversial. Even where bans have been enacted, many drivers simply ignore the law. This leaves lawmakers and others with an open question on the matter of cell phones and safe driving practices.

Facts

- According to the April 20, 2010, Pew Research Center report *Teens and Mobile Phones*, 48 percent of all teens aged 12 to 17 say they have been in a car when the driver was texting; the same percentage say they have been in a car when the driver used a cell phone in a way that put themselves or others in danger.

- Twenty-nine states ban all kinds of cellular phone use by novice drivers. Novice drivers are generally categorized as those below 18 or those who only have a learner's permit or provisional license.

- According to a 2009 study conducted by the California Highway Patrol, text-messaging drivers spend up to approximately 400 percent less time looking at the road compared with nontexting drivers.

- According to the University of North Texas Health Science Center, the percentage of all traffic deaths caused by distracted driving rose from 11 percent in 1999 to 16 percent in 2008.

- Fifteen states and DC use social networking sites like Twitter and Facebook to promote anti-distracted-driving messages.

How Do Cell Phones Impact Youth?

Thirteen-year-old Hope Witsell was an average middle school student when she sent a topless photo of herself to a boy she was romantically interested in. A third party intercepted the photo and forwarded it on to other students. Eventually, the photo was spread throughout not only Hope's school but other neighboring schools as well. Other students at Hope's school made her life a living hell, hurling insults her way and continually harassing her about the photo. During the summer, Hope faced grounding by her parents and, when school began again, she was suspended for the first week of school and told she could no longer be the adviser for the Future Farmers of America (FFA) program, an activity she loved.

While meeting with Hope, school counselors noticed evidence that she had been cutting herself. They had her sign a "no-harm contract," meant to have her agree to tell an adult if and when she felt like hurting herself. The school neglected to inform the parents of Hope's self-destructive actions. The following day Hope wrote in her journal, "I am done for sure now. I can feel it in my stomach. I'm going to try and strangle myself. I hope it works."[45] No longer able to cope with the pressures she felt were crushing her, Hope committed suicide by hanging herself with a scarf from the canopy of her bed. Hope's tragic story illustrates how cell phones have allowed young boys and girls to make mistakes that have repercussions far beyond their coping capabilities.

An Essential Tool for Teens

Cell phone use is ubiquitous among American teenagers. Seventy-five percent of American teens aged 12 to 17 have a cell phone. Seventy-two percent of all teens send text messages. Teens who have a cell phone consider it an essential part of their lives, using it to play music, videos, and games and to call and text friends. According to one study by Context, a Baltimore research group of anthropologists who study consumer trends, being without a cell phone constitutes social suicide for a teen. Chief anthropologist of the study Robbie Blinkoff concludes, "If you don't have the technology, you are not part of the class. . . . If you are not a name or number on my phone book, then you are not on my radar screen."[46] Blinkoff noted that a digital divide separated teens with cell phones from teens without. He also noticed that groups of teens would often interact more with their cell phones when in a group than with each other. Rather than being considered rude, teens viewed it as a group activity like hanging out at the mall or going to a movie with friends. Blinkoff concluded that this was a new form of socializing.

While cell phones may represent a dividing line between being "in" or "out," one finding by a Pew Research study suggests that among teens from poor families, the cell phone provides at least one source for narrowing the digital divide: access to the Internet. The study found that many students from poor families had cell phones even if they had no computer at home. Minorities were more likely to use their phones to access the Internet, including 44 percent of black teens and 35 percent of Hispanic teens.

Health Effects for Teens

Cell phone use among younger and younger users is just beginning to be researched, and the findings change as more studies become available. A 2008 study by Gay Badre of Sahlgren's Academy in Sweden studied 21 subjects between the ages of 14 and 20 to monitor the effects of cell phone use on sleep patterns. One group made fewer than five calls or texts a day, while the other group made more than 15 calls or texts a day. Teens who talked and texted a lot more were more restless, careless, drank stimulating drinks, and had difficulty in either falling asleep or staying asleep than those who hardly used their cell phones. Badre remarked, "Addiction to [the] cell phone is becoming common. Youngsters feel a group pressure to remain inter-connected and reachable round the clock. . . . There seems to be a connection between intensive use of cell phones and health compromising behavior such as smoking, snuffing, and use of alcohol."[47] It appears that other negative physiological effects are also beginning to reveal themselves as cell phone use among the young continues to be studied.

Cell Phones and the Young Brain

Recent studies have proven that the human brain is exposed to radiation when using a cell phone. Scientists are concerned that this radiation may affect teens and children differently than adults, since young people have thinner skulls and their brains have not finished developing. Preliminary studies find that the dangers range from unsettling to deeply disturbing. Leif Salford claims "the voluntary exposure of the brain to microwaves from hand-held mobile phones is the largest human biological experiment ever."[48] Om Gandhi, a leading scientist and professor of electrical engineering at the University of Utah contends that young children under 10 could absorb radiation across their entire brains. One of the most disturbing findings was a study published in the May 2008 issue of the *International Journal of Oncology* from Sweden. Looking at the general population, the researchers saw a fivefold increase in certain cancerous brain tumors after 10 years of exposure to cell phone radiation. Since tumors can take more than 10 years to develop, these findings seem particularly disturbing regarding young users. As Devra Davis, president and founder of the Environmental Health Trust of San Francisco remarked, "Because children's skulls, brains and bodies are thinner and more vulnerable, we put them in bicycle helmets and car seats. We need to take parallel steps to protect them. . . . We need to protect their developing brains and bodies from exposure to a sea of radiofrequency radiation whose full impact cannot be gauged at this time."[49]

"There seems to be a connection between intensive use of cell phones and health compromising behavior such as smoking, snuffing, and use of alcohol."[47]

— Gay Badre, researcher at Sahlgren's Academy in Sweden.

While many medical professionals are recommending caution when it comes to teens and cell phones—including having teens only use a hands-free device or ear phones to keep cell phones away from their brains—definitive answers are still in the future.

Cell Phones in the Schools

While young peoples' health may be at stake when it comes to cell phone use, the devices are also affecting their education. School policy on cell phones ranges from a complete ban on all phones to schools allowing students to use their phones during breaks and before and after school. While most schools have rules regarding phone use, most students ignore these rules, according to a Pew Research study. Nearly 65 percent of teens at "no phone" schools bring their cell phone to school every day anyway. Four in five (81 percent) of teens at schools where phones are allowed bring a phone to school every day.

> "The voluntary exposure of the brain to microwaves from hand-held mobile phones is the largest human biological experiment ever."[48]
>
> — Leif Salford, head of cell phone radiation research at Lund University in Sweden.

Cell phones do prove a distraction in school. Administrators and teachers complain that students interrupt class time by using their phones to sneak text messages. One middle school boy in the Pew study said, "When I'm in class I just see people pull out their phone and try to be sneaky, and get past the teachers and try texting and stuff."[50] The Pew study also noted that the number of teens who used a cell phone during class time seemed unrelated to the policies of the school. Roughly one-third of students attempted use of the phones, one-third claimed never to do so, and one-third claimed to do so only occasionally.

One tempting use of cell phones is for cheating. In a study by Common Sense Media in 2010, two-thirds of all students questioned said they knew of classmates using their phones to cheat on tests. More than one-third admitted to doing it themselves. Most students who use their phones to cheat on tests do so in one of several ways. A student may text an answer to another student, take pictures of an exam to give to another student, take pictures of textbook pages to use in the exam, or access the Internet for answers during the exam. Other features of the phone, such as the calculator, can also be used to cheat in class. Another high school

Students at an Alabama high school catch up on their cell phone calls during a break between classes. Some districts have banned cell phones from schools while others have adopted rules on when and where cell phone use is permitted.

student quoted in the Pew study said, "I have heard of people taking pictures of the textbook—a section where they didn't know anything—and then they'll zoom in and take a photo. It's kind of like a sheet of notes, but it is on your phone. It's not like you have to pull out a piece of paper and unfold it and make a lot of noise. It is easier."[51]

Cell Phones as Educational Tools

Many teachers agree that while the cell phone makes it easier to cheat, they have always dealt with the issue of cheating and must continue to be vigilant to catch the cheaters. Some educators have gone even further by allowing students to use the phones in assignments. Daniel A. Domenech, of the American Association

Girls Use Cell Phone Technologies More

According to a 2010 study by Pew Research, girls are more likely than boys to use cell phone technology, including both voice mail and texting. Girls on average send and receive 80 texts a day, compared with boys, who send and receive an average of 30. More than half (59 percent) of girls called their friends just to say hello and chat.

Girls are also more likely to be familiar with more complicated phone technologies to keep in touch. For example, girls are more likely to understand and use the conference call function of the cell phone and are more familiar with this technology than their parents.

of School Administrators, argues that "handheld devices like cell phones, iPhones, BlackBerrys and ITouch are beginning to offer applications that enhance classroom learning by engaging kids to use tools they are constantly using anyway."[52]

Corporations and educators are getting on board to engage students by using their cell phones as educational tools. Abilene Christian University handed out Apple iPhones to two-thirds of its freshman class in 2008. Students used them to brainstorm ideas and get virtual handouts and podcasts during class. Software makers have caught the trend and have been pumping out education applications, from math programs to telling time to learning languages. Administrators see the use of phones as a cheaper alternative to computers with just as many tools, including Internet access, e-mail, and educational software.

One example comes from educator Liz Colb, who teaches teachers and is the author of the book *Toys to Tools: Connecting Student Cell Phones to Education*. She encourages teachers to try cell phone exercises with their students. She has taught teachers how to find, download, and help students use cell phone apps to work on such activities as creating raps about math, teaching foreign languages, and recording podcasts.

Bullying

While schools are finding ways to use cell phones in a positive way, teens' use of the devices is mixed. While some teens use cell phones to share photos, videos, and other materials instantly with a wide audience, others have used it as a way to bully and harass. Cell phone bullying includes leaving hateful or threatening voice mails, text messages, or sending compromising pictures of the victim directly to his or her cell phone or to someone else's cell phone. If the victim has a smart phone with a Facebook or other social networking site application, he or she can be further harassed by hurtful posts. According to the Pew Research Center, 26 percent of teen cell phone users reported having been harassed by someone via cell phone.

Text bullying allows for a teen to send harassing messages without ever having to face his or her victim or leave the privacy of his or her own home, sometimes making the sender more uninhibited and therefore more vicious. One teen girl said, "They can make chain letters about people, take pictures of people doing stuff in class, like picking their nose or sleeping or something, and add really mean soundtracks in the background, write really mean stuff in there, send it to everyone in their phonebook."[53] Text messaging is harder for a victim to combat and harder to ignore than the Internet or in-person harassment. A victim can be available for text bullying 24 hours a day.

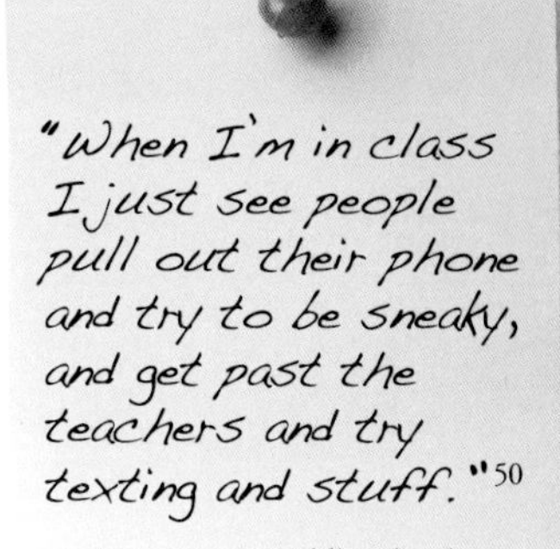

The most vulnerable victims of text bullying and other types of cyberbullying may be middle schoolers, but the problem also applies to high school students. Because rejection from peers can seem overwhelmingly hurtful, middle and high school students, already coping with many emotional and hormonal changes, can be pushed to the edge when they are bullied via cell phone and on the Internet. In January 2010, for example 15-year-old Phoebe Prince committed suicide by hanging after being bullied with text messages, Facebook posts, and in-school harassment.

Schools and parents are often hamstrung by legalities and senders' ability to hide their tracks while texting. Schools cannot search a cell phone because of the privacy of students, and harassment that is done outside of school is usually beyond the school's purview. In addition, most schools cannot interfere with students' right to free speech. One principal at Governor Livingston High School in Berkeley Heights, New Jersey, says that the school can only search a cell phone when a student says that someone has sent him or her bullying messages.

Sexting

Sexting involves sending, receiving, or forwarding pornographic photos, e-mails, or text messages. It is the latest flirtation method among teens and a phenomenon that continues to gain in popularity. In a 2010 survey posted by the National Campaign to Prevent Teen and Unplanned Pregnancy, a private organization, 20 percent of all teens have posted or sent nude or seminude pictures of themselves; 11 percent of these were young teen girls aged 13 to 16. The number who sent sexually explicit messages was even higher; 39 percent of all teenagers, 37 percent of teenage girls, and 40 percent of teen boys admit to the activity. Fifty-one percent of teen girls say they felt pressure from boys to send sexually suggestive or explicit messages or pictures. Over 60 percent of both teen boys and girls said they did so for fun or to be flirtatious.

Studies suggest that teens find sexting harmless. In 2011 the *New York Times* interviewed teenagers individually and in focus groups about texting and found that students' were very matter-of-fact about the subject. One interviewee, Kathy, 17, said, "At my school, if you like a boy and you want to get his attention, you know what you have to do. When I was with my last boyfriend I refused to sext and I would go through his iPod and find pictures of girls' breasts. . . . And they were all girls in my school."[54] Other students suggested that sexting was a safe alternative to having sexual contact. Some psychologists argue that sexting is a natural consequence of a society that glorifies and normalizes sexual contact and sexually explicit images.

Girls, who sext more often than boys, see it as a special gift to a boyfriend or boy that they are interested in. Rachel, 18, said, "A girl thinks, 'I know I've been warned against it, but this is something I want to share with my boyfriend, and he's different.'"[55] Un-

fortunately, many girls who sext may be fooling themselves about the privacy of the act. William, 18, said, "If a girl sent me a picture of her boobs, well, obviously I'd like to show it to some friends."[56]

Some educators have embraced cell phones as a teaching tool. Abilene Christian University handed out Apple iPhones (newer versions of which are pictured) to a large chunk of its freshman class in 2008. Students used them to brainstorm ideas and get virtual handouts and podcasts during classes.

Privacy Is an Illusion

Other publicized incidents make it clear that "private" is the last thing sexually explicit images remain. In 2011 the *New York Times* reported on an incident involving Margarite, an eighth grader who sent a full-length nude picture to her boyfriend, Isaiah. After they broke up a few weeks later, Isaiah's new girlfriend found the picture, added the caption "Ho Alert! If you think this girl is a whore, then text this to all your friends."[57] The girl then forwarded the picture to all of her contacts, and those contacts forwarded it to more. The photo went "viral," the term used when a cyber text becomes wildly popular. Suddenly, Margarite's naked body was on cell phones everywhere.

The county prosecutor of Lacey, Washington, where the incident took place, decided to charge three students with the dissemination of child pornography. The charge, a Class C felony, would mean that the students would be listed as registered sex offenders. When nude pictures of underage boys or girls are sent or received via cell phone, child pornography laws can apply. The prosecutor eventually decided to charge the three students who had helped disseminate Margarite's photo with telephone harassment, allowing them to have the charges dismissed after community service.

Sexting Laws

Many states are in the process of attempting to define sexting and developing laws against it. Most prosecutors, school administrators, and parents agree that charging minors under current child pornography laws seems overly harsh. Some states, such as Nebraska, punish those who forward the message, not the creator, arguing that those who forward the message are the ones who create the harassment. Other states are proposing misdemeanor laws for those who send or forward sexts.

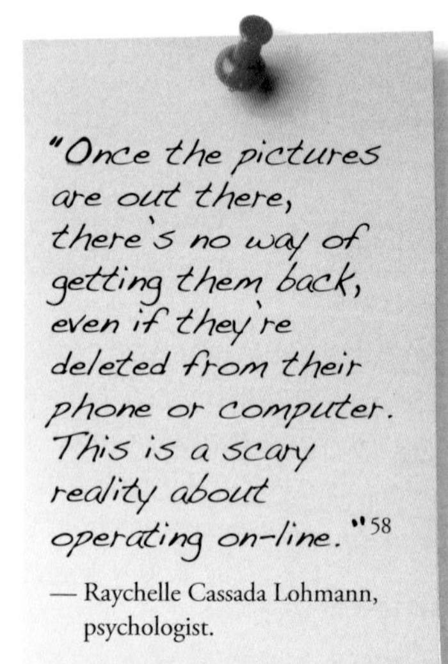

— Raychelle Cassada Lohmann, psychologist.

In 2009 District Attorney George Skumanick Jr. of Tunkhannock, Pennsylvania, forced students who had texted nude photos of themselves to enlist in an education program or face prosecution for child pornography. When students refused to enlist in the program, Skumanick moved forward with the child pornography charges. Skumanick was prevented from prosecuting the case when the parents of some of the students and the ACLU sued Skumanic in federal court alleging he violated the freedom of expression rights of the students.

While legislators work out the details of how to cope with the legalities of texting, parents worry that the impulsive and emotional nature of their teen's actions could mean that sexually explicit photos of their child could remain on the Internet for years. Those photos could be viewed by anyone accessing the Internet and may plague their child throughout his or her adult life. In *Psychology Today*, psychologist Raychelle Cassada Lohmann says that teens should be told that "once the pictures are out there, there's no way of getting them back, even if they're deleted from their phone or computer. This is a scary reality about operating on-line. Let your teens know that www not only stands for 'World Wide Web' it also stands for 'Whole World's Watching.'"[58]

Before cell phones, teens' impulsivity could lead them into trouble, but rarely did that trouble follow them into their adult lives. Now teens are making decisions at a young age that may prove to have disturbing consequences. All parties, including parents, legislators, schools, and the students themselves, are in uncharted territory that may take decades to sort out.

Facts

- **According to the Pew Research Center, sexting occurs in three scenarios: Exchanges of images between romantic partners; exchanges of images between partners that are then forwarded to others; and exchanges of images for the purpose of attempting to become romantic partners.**
- **According to a National Campaign to Prevent Teen and Unplanned Pregnancy survey, 20 percent of all teens have sent or posted a sexually revealing photo or video.**
- **Students who send or receive sexual pictures or content on their cell phones may be subject to prosecution for child pornography.**
- **According to a survey by the National Campaign, 71 percent of teen girls and 67 percent of teen boys who have sent or posted sexually suggestive content say they have sent or posted this content to a boyfriend or girlfriend.**
- **According to the Pew Research Center, 26 percent of teen cell phone users reported having been harassed by someone else through their cell phone.**

- **According to the Pew Research Center, two out of every five youths in the United States between the ages of 8 and 18 own a cell phone.**

- **According to a study by the University of Granada, 40 percent of young adults admit using their mobile phones more than four hours a day.**

- **According to a CommonSense Media survey, 25 percent of American students admitted to text messaging a friend to get an answer to a question during a quiz.**

- **Children crossing streets while using cell phones were 43 percent more likely to have a close call or actually be struck by a passing vehicle, according to 2008 research by the University of Alabama at Birmingham and published in *Pediatrics*.**

- **According to the April 20, 2010, Pew Research Center report *Teens and Mobile Phones*, half (52 percent) of cell-owning teens aged 16 and 17 say they have talked on a cell phone while driving.**

Related Organizations and Websites

Cellular Telecommunications & Internet Association (CTIA)
1400 16th St. NW, Suite 600
Washington, DC 20036
phone: (202) 736-3200
fax: (202) 785-0721
website: www.ctia.org

CTIA is an international nonprofit organization that represents the wireless communications industry. Members include wireless carriers as well as providers and manufacturers of wireless data services and products. CTIA supports a variety of public service campaigns to promote the safe use of cell phones, including "On the Road, Off the Phone."

Environmental Health Trust (EHT)
PO Box 58
Teton Village, WY 83025
e-mail: info@ehtrust.org
website: www.environmentalhealthtrust.org

The EHT seeks to educate individuals and health professionals about controllable environmental health risks and policy changes needed to reduce those risks. Current multimedia projects include a campaign promoting safer cell phone practices and calling for greater innovation on behalf of the cell phone industry to make cell phones safer.

Environmental Working Group (EWG)

1436 U St. NW, Suite 100
Washington, DC 20009
phone: (202) 667-6982
website: www.ewg.org

The EWG is a nonprofit organization of scientists, engineers, policy experts, lawyers, and others who seek to protect public health by disseminating published reports and analyses of a variety of potential health threats, including cell phones. The group's *Guide to Cell Phone Radiation* provides the radiation emissions of more than 1,000 cell phones currently on the market.

Federal Communications Commission (FCC)

445 12th St. SW
Washington, DC 20554
phone: (888) 225-5322
fax: (866) 418-0232
website: www.fcc.gov

The FCC is an independent government agency overseen by Congress that regulates communications by radio, television, wire, satellite, and cable in the United States. The FCC certifies that all phones sold in the United States comply with FCC guidelines on RFR exposure. The agency also regulates cell phone base stations.

Governors Highway Safety Association (GHSA)

444 N. Capitol St. NW, Suite 722
Washington, DC 20001-1534
phone: (202) 789-0942
fax: (202) 789-0946
e-mail: headquarters@ghsa.org
website: www.ghsa.org

The GHSA implements programs addressing behavioral issues affecting highway safety, including distracted driving, the most common of which it believes is cell phone use and texting. The GHSA report *Curbing Distracted Driving: 2010 Survey of State Safety Programs* advocates ways for states to respond to the problem and promote safer driving practices.

International Agency for Research on Cancer (IARC)

150 Cours Albert Thomas
69372 Lyon CEDEX 08
France
phone: 33 04 72 73 84 85
website: www.iarc.fr

The IARC, part of the World Health Organization, promotes international collaboration in research on the causes of cancer and how to develop strategies for cancer prevention and control. IARC's research activities include a large international collaborative study on cell phone use and the risk of brain tumors, acoustic neuromas, and salivary gland tumors.

National Cancer Institute (NCI)

6116 Executive Blvd., Suite 300
Bethesda, MD 20892-8322
phone: (800) 422-6237
e-mail: cancergovstaff@mail.nih.gov
website: www.cancer.gov

The NCI is part of the National Institutes of Health. As the federal government's principal cancer agency, NCI conducts and supports research into the causes and treatment of cancer, maintains a cancer library, and disseminates information on a variety of topics, including health issues associated with RF radiation from cell phones.

National Crime Prevention Council (NCPC)

2001 Jefferson Davis Hwy., Suite 901
Arlington, VA 22202-4801
phone: (202) 466-6272
fax: (202) 296-1356
website: www.ncpc.org

The NCPC implements a variety of programs to keep individuals and communities safe from crime. The council disseminates publications and teaching materials on topics, including cell phone safety, sexting, cyberbullying, and other issues that affect youth.

NCI programs include Teens, Crime, and the Community and Youth Outreach for Victim Assistance.

National Highway Traffic Safety Administration (NHTSA)

1200 New Jersey Ave. SE, West Bldg.
Washington, DC 20590
phone: (888) 327-4236
website: www.nhtsa.gov

The mission of NHTSA, part of the US Department of Transportation, is to prevent traffic-related injuries and deaths. NHTSA disseminates information to promote a greater understanding of the issue of distracted driving and works with a variety of private and public groups and advocacy groups to end the practice of using cell phones while driving.

National Research Center for Women and Families (NRC)

1001 Connecticut Ave. NW, Suite 1100
Washington, DC 20036
phone: (202) 223-4000
website: www.center4research.org

The NRC is a nonprofit organization that promotes health and safety by encouraging the implementation of effective policies, programs, and medical treatments based on research. The group disseminates information on a variety of cell phone topics, such as physical health implications and the social impact of cell phones.

Radiation Research Trust (RRT)

Chetwode House
Leicester Rd.
Melton Mowbray
Leicestershire LE13 1GA
UK
phone: 44 1664 414500
e-mail: eileen@radiationresearch.org
website: www.radiationresearch.org

The RRT disseminates information about electromagnetic radiation and health to the public and media. As part of its campaign to raise awareness about cell phone safety, the organization lobbies the UK government and the European Union to fund and prioritize research and campaigns for policy changes based on research findings.

US Food and Drug Administration (FDA)

10903 New Hampshire Ave.
Silver Spring, MD 20993
phone: (888) 463-6332
website: www.fda.gov

The FDA, a part of the Department of Health and Human Services, belongs to the Radiofrequency Interagency Work Group, which is responsible for RF safety. The FDA has the authority to take action if cell phones are shown to emit RF radiation at harmful levels. Its website provides information about cell phone safety and current research.

Additional Reading

Books

Carleigh Cooper, *Cell Phones and the Dark Deception: Find Out What You're Not Being Told and Why*. Grandville, MI: Premier Advantage, 2009.

Devra Davis, *Disconnect: The Truth About Cell Phone Radiation, What the Industry Has Done to Hide It, and How to Protect Your Family*. New York: Dutton, 2010.

Ann Louise Gittleman, *Zapped: Why Your Cell Phone Shouldn't Be Your Alarm Clock and 1,268 Ways to Outsmart the Hazards of Electronic Pollution*. New York: HarperOne, 2010.

Uwe Hansmann and Lothar Merk, *Pervasive Computing: The Mobile World*. New York: Springer, 2011.

Sameer Hinduja and Justin Patchin, *Bullying Beyond the Schoolyard: Preventing and Responding to Cyberbullying*. Thousand Oaks, CA: Sage, 2009.

Guy Klemens, *The Cellphone: The History and Technology of the Gadget That Changed the World*. Jefferson, NC: McFarland, 2010.

Rich Ling and Jonathan Donner, *Mobile Phones and Mobile Communication*. Stafford, Australia: Polity, 2009.

Periodicals

Michael Austin, "Texting While Driving: How Dangerous Is It?," *Car and Driver*, June 2009.

Sharon Begley, "Will This Phone Kill You?," *Newsweek*, August 5, 2010.

Frank Bures, "Can You Hear Us Now?," *Utne Reader*, March/April 2011.

Marlene Cimons, "Cell Phones as Community Observers," *U.S. News & World Report*, May 4, 2011.

Noam Cohen, "It's Tracking Your Every Move and You May Not Even Know It," *New York Times*, March 26, 2011.

Larry Copeland, "Driver Phone Bans' Impact Doubted," *USA Today*, January 29, 2010.

Economist, "Think Before You Speak: Distracted Driving Is the New Drunk Driving," April 14, 2011.

Nic Fleming, "Largest Ever Cellphone Cancer Study Is Inconclusive," *New Scientist*, May 22, 2010.

Ashley Halsey III, "28 Percent of Accidents Involve Talking, Texting on Cellphones," *Washington Post*, January 13, 2010.

Jerry Hirsch, "Teens, Driving, and Texting Are a Bad Mix," *Los Angeles Times*, August 2, 2010.

Jan Hoffman, "States Struggle with Minors' Sexting," *New York Times*, March 26, 2011.

Geoffrey Lean, "Mobile Phone Use 'Raises Children's Risk of Brain Cancer Fivefold,'" *Independent* (London), September 21, 2008.

Judith Levine, "What's the Matter with Teen Sexting?," *American Prospect*, February 2, 2009.

Maclean's, "Cellphone Bans Aren't Making the Roads Any Safer," August 30, 2010.

Kate Murphy, "Cellphone Radiation May Alter Your Brain. Let's Talk," *New York Times*, March 30, 2011.

Donna St. George, "Study of Teen Cellphone Use Reinforces Impression That They're Always Using Them," *Washington Post*, April 20, 2010.

Bryan Walsh, "Cell Phone Safety," *Time*, March 15, 2010.

Internet Sources

Devra Davis, "Beyond Brain Cancer: Other Possible Dangers of Cell Phones," *Huffington Post*, June 15, 2011. www.huffingtonpost.com.

Governors Highway Safety Association, "Cell Phone and Texting Laws," June 2011. www.ghsa.org.

National School Safety and Security Services, "Cell Phones and Text Messaging in Schools," 2010. www.schoolsecurity.org.

Pew Research Center, "Teens and Mobile Phones," April 20, 2010. www.pewinternet.org.

Dan Shapley, "Cell Phones May Be Unsafe for Teens," The Daily Green, June 28, 2011. www.thedailygreen.com.

Source Notes

Introduction: A Ubiquitous Technology

1. Quoted in News-Medical.net, "Mobile Phone Addiction in Teenagers May Cause Severe Psychological Disorders," February 27, 2007. www.news-medical.net.

Chapter One: What Are the Origins of the Cell Phone Debate?

2. Guy Klemens, *The Cellphone: The History and Technology of the Gadget That Changed the World.* Jefferson, NC: McFarland, 2011, p. 45.
3. Quoted in Devra Davis, *Disconnect: The Truth About Cell Phone Radiation, What the Industry Has Done to Hide It, and How to Protect Your Family*. New York: Dutton, 2010, p. 13.
4. Klemens, *The Cellphone*, p. 1.
5. Klemens, *The Cellphone*, p. 1.
6. Davis, *Disconnect*, pp. 2–3.
7. Quoted in Davis, *Disconnect,* p. 44.
8. Quoted in Davis, *Disconnect*, p. 44.
9. National Cancer Institute, "Cell Phones and Cancer Risk Fact Sheet," June 2011. www.cancer.gov.
10. Davis, *Disconnect*, p. 75.
11. Quoted in Randall Stross, "Should You Be Snuggling with Your Cellphone?," *New York Times*, November 13, 2010. www.nytimes.com.
12. National Cancer Institute, "Cell Phones and Cancer Risk Fact Sheet."
13. Gerard Goggin, *Cell Phone Culture: Mobile Technology in Everyday Life.* New York: Routledge, 2006, p. 2.

14. International Telecommunication Union, "The Social Impact of Mobile Telephony," 2004. www.itu.int.

15. International Telecommunication Union, "The Social Impact of Mobile Telephony."

16. Klemens, *The Cellphone*, p.159.

Chapter Two: Does Cell Phone Use Lead to Cancer?

17. Klemens, *The Cellphone*, p. 143.

18. Klemens, *The Cellphone*, p. 143.

19. Quoted in Danielle Dellorto, "WHO: Cell Phone Use Can Increase Possible Cancer Risk," CNN.com, May 31, 2011. www.cnn.com.

20. Quoted in Tara Parker-Pope, "Cellphones May Cause Cancer," *New York Times*, May 31, 2011. www.nytimes.com.

21. Michael Shermer, "Can You Hear Me Now?," *Scientific American*, October 2010. www.scientificamerican.com.

22. Eric Swanson, "Stop Freaking Out About Cell Phones!," *Pittsburgh Post-Gazette*, August 3, 2008. www.postgazette.com.

23. Quoted in Tara Parker-Pope, "Piercing the Fog Around Cellphones and Cancer," *New York Times*, June 6, 2011. www.nytimes.com.

24. Ronald Herberman, "Tumors and Cell Phone Use: What the Science Says," statement to the Domestic Policy Subcommittee, Oversight and Government Reform Committee, September 25, 2008, p. 5.

25. Quoted in Joe Fahy, "Cancer Chief Sees Cell Phone Risks," *Pittsburgh Post-Gazette*, July 23, 2008. www.postgazette.com.

26. Quoted in Swanson, "Stop Freaking Out About Cell Phones!"

27. Quoted in Dellorto, "WHO: Cell Phone Use Can Increase Possible Cancer Risk."

28. Klemens, *The Cellphone*, p. 154.

29. Devra Davis, "Cell Phones and Brain Cancer: The Real Story," *Huffington Post*, May 22, 2010. www.huffingtonpost.com.

30. Klemens, *The Cellphone*, p. 144.

Chapter Three: Do Cell Phones Cause Other Health Problems?

31. Quoted in Dellorto, "WHO: Cell Phone Use Can Increase Possible Cancer Risk."

32. Davis, *Disconnect*, p. 130.

33. Ashok Agarwal, "Cell Phone Radiation Degrades Semen Quality," *Urology News*, 2008. www.clevelandclinic.org.

34. Agarwal, "Cell Phone Radiation Degrades Semen Quality."

35. Quoted in *Daily Mail Reporter*, "Talking on a Mobile Phone 'May Lower Male Fertility,'" MailOnline, May 20, 2011. www.dailymail.co.uk.

36. Quoted in Davis, *Disconnect*, p. 65.

37. Quoted in *ScienceDaily*, "Mobile Phones Affect Memory in Laboratory Animals, Swedish Study Finds," December 5, 2008. www.sciencedaily.com.

38. Quoted in Mary Brophy Marcus, "Cellphone Use Affects Brain Activity," *Tucson (AZ) Citizen*, February 22, 2011. www.tucsoncitizen.com.

39. Quoted in Kate Murphy, "Cellphone Radiation May Alter Your Brain. Let's Talk," *New York Times*, March 30, 2011. www.nytimes.com.

Chapter Four: How Has Cell Phone Use Affected Driving?

40. Quoted in Daniella Guzman, "Mom Warns About Cell Phone Use While Driving," Click2Houston.com, October 21, 2010. www.click2houston.com.

41. David L. Strayer et al., "A Comparison of the Cell Phone Driver and the Drunk Driver," *Human Factors*, Summer 2006, p. 388.

42. Quoted in Gilbert Cruz and Kristi Oloffson, "Driving Us to Distraction," *Time International*, August 31, 2009. search.ebscohost.com.

43. *News from Vanderbilt University*, "Neural Bottleneck Found That Thwarts Multi-Tasking," January 18, 2007. sitemason.vanderbilt.edu.

44. Safe Driving Systems, "Texting While Driving: Danger Eliminated with New Technology," March 21, 2010. www.safedrivingsystems.com.

Chapter Five: How Do Cell Phones Impact Youth?

45. Quoted in Michael Inbar, "Sexting Bullying Cited in Teen's Suicide," Today.com, December 2, 2009.

46. Quoted in Elisa Batista, "She's Gotta Have It: Cell Phone," *Wired*, May 16, 2003. www.wired.com.

47. Quoted in *Science Daily*, "Excessive Mobile Phone Use Affects Sleep in Teens, Study Finds," June 8, 2008. www.sciencedaily.com.

48. Quoted in *Childhood Genius Internet Magazine*, "Developing the Child Brain," 2011. internationalparentingassociation.org.

49. Devra Davis, "Beyond Brain Cancer: Other Possible Dangers of Cell Phones," *Huffington Post*, June 15, 2011. www.huffingtonpost.com.

50. Quoted in PewResearchCenter, "Teens and Mobile Phones," April 20, 2010. pewinternet.org.

51. Quoted in PewResearchCenter, "Teens and Mobile Phones."

52. Quoted in Alex Johnson, "Some Schools Rethink Bans on Cell Phones," MSNBC, February 3, 2010. www.msnbc.com.

53. Quoted in Kathy Brock, "Text Bullying," ABC News, May 9, 2008. http://abclocal.go.com.

54. Quoted in *New York Times*, "What They're Saying About Sexting," March 26, 2011. www.nytimes.com.

55. Quoted in *New York Times*, "What They're Saying About Sexting."

56. Quoted in *New York Times*, "What They're Saying About Sexting."
57. Quoted in Jan Hoffman, "A Girl's Nude Photo, and Altered Lives," *New York Times*, March 26, 2011. www.nytimes.com.
58. Raychelle Cassada Lohmann, "Sexting Teens," *Psychology Today*, March 30, 2011. www.psychologytoday.com.

Index

Picture Credits

Cover: iStockphoto.com

AP Images: 13, 17, 23, 40, 55, 59, 67, 71

Mauro Fermariello/Science Photo Library: 37

GEC Research/Hammersmith Hospital Medical School/Science Photo Library: 31

Steve Gschmeissner/Science Photo Library: 49

C.J Guerin, PhD, MRC Toxicology Unit/Science Photo Library: 45

Thinkstock/iStockphoto.com: 9

About the Authors

Bonnie Szumski has been an editor and author of nonfiction books for over 25 years. Jill Karson has been a writer and editor of nonfiction books for young adults for 15 years.